STYLING

CLASSIC HAIRSTYLE DESIGNS

黑光造型

经典造型设计专业教程

黑光造型 编著

人民邮电出版社

北京

图书在版编目（CIP）数据

经典造型设计专业教程 / 黑光造型编著. -- 北京 :
人民邮电出版社, 2015.1
（黑光造型）
ISBN 978-7-115-36996-3

Ⅰ. ①经… Ⅱ. ①黑… Ⅲ. ①发型－设计－教材
Ⅳ. ①TS974.21

中国版本图书馆CIP数据核字(2014)第293132号

内 容 提 要

本书分为经典白纱造型设计解析、经典晚礼造型设计解析和特色服装造型设计解析三个部分，其中包括大量的实例教程，旨在分步解析技巧要领，并分析当今流行的造型风格。书中的作品形式丰富，风格唯美，文字表述简明扼要，通俗易懂，是一本不可多得的参考资料。本书与《黑光造型 创意造型设计佳作赏析》为一套图书，后者在本书的基础上注重提高，旨在通过顶尖造型实例作品，解析创意灵感，为创意人士提供灵感参悟和实现方法。

本书既适合造型设计专业人士研修学习，也便于造型设计爱好者和摄影师借鉴与参考，以提高艺术水准和造型实力。

◆ 编　　著　黑光造型
责任编辑　赵　迟
责任印制　程彦红

◆ 人民邮电出版社出版发行　　北京市丰台区成寿寺路 11 号
邮编　100164　　电子邮件　315@ptpress.com.cn
网址　http://www.ptpress.com.cn
北京盛通印刷股份有限公司印刷

◆ 开本：880×1092　1/16
印张：16.5
字数：659 千字　　2015 年 1 月第 1 版
印数：1－3 000 册　　2015 年 1 月北京第 1 次印刷

定价：98.00 元（附光盘）

读者服务热线：(010)81055410　印装质量热线：(010)81055316
反盗版热线：(010)81055315

序

本书的问世是业界的一件盛事！

在业界一直流传一句让我感到骄傲的话：“黑光是出人才的地方！”本书就是那些勤劳的、整日忙碌于教学第一线的众多黑光名师和美丽的模特们辛勤劳动的结晶……

本书定位于全方位、多层次、多角度，就像我们身边这个多姿多彩的世界一样，黑光向来主张艺术多元化。与黑光以前出版的作品集《黑光造型宝典》《新娘100%》相同，本书也不是哪一位老师的个人作品集，而是众多黑光老师用心血和智慧共同完成的一整套的实用化妆造型教程与作品集萃。其风格千变万化，是化妆造型的一场饕餮盛宴！本书不仅展现了黑光众多名师化妆造型作品的艺术魅力，更是与时尚的一场心灵对话。需要说明的是，书中的很多作品并非刻意之作，而是黑光老师们平日授课时信手拈来的作品，这更能体现黑光老师们深厚的艺术造诣。

《黑光造型 经典造型设计专业教程》与《黑光造型 创意造型设计佳作赏析》是一套图书。前者旨在分步解析技巧要领，与读者分享当今流行的造型设计的创作方法；后者旨在通过解析顶尖造型作品的创意思路，为创意人士提供灵感参悟和实现方法。本套图书整体上追求简明扼要、通俗易懂，既适合造型设计专业人士研修学习，也便于造型设计爱好者、化妆师、造型师、摄影师借鉴与参考，以提高艺术水平和造型软实力。

时代在变迁，过分雕琢的、匠气的化妆造型已经不适应现在的需求。删繁就简、贴近生活、真切、自然、清新、时尚、与国际接轨、实用性强的化妆造型风格才是流行的主旋律，这也是本套图书始终贯彻的一个宗旨。

本套图书采用中英文对照，教程部分的案例配以步骤图，力求做到图文并茂，让读者一看就懂，一学就会。本套图书采用了教材的形式，具有极强的实用性，对于自学者来说也是难得的好教材。

希望读者在充分利用好黑光系列丛书的同时，能够融入到黑光教学课堂中参加学习与交流，我们将努力使读者得到更大的提升。

本书是黑光老师们在完成繁忙的教学工作后利用业余时间编纂的，旨在进行技术交流。如有不完美的地方，深望同行予以斧正，让我们一起为中国的化妆造型事业做出更大的贡献！

北京黑光教育咨询有限公司董事长 刘伟庆

黑光360造型STYLING

目录

STYLING TOOLS AND USES

造型工具及用途

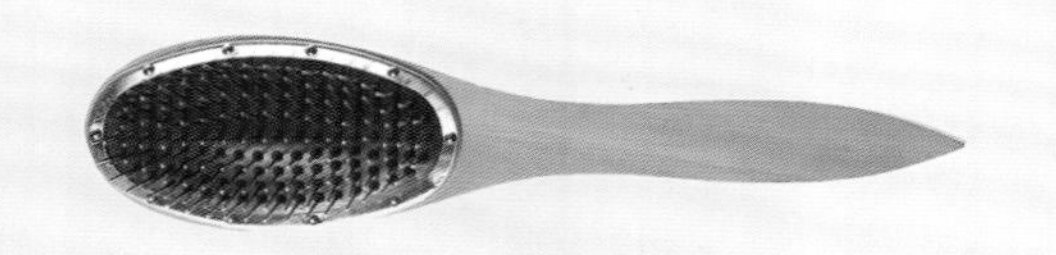

钢齿梳 Steel-tooth comb

用于梳理假发，不易使假发纤维变形。

It is used to comb wig without deforming the wig fiber.

扁平鬃毛梳 Flat bristle comb

有着一层柔软的鬃毛的扁平梳，适合在头发需要营造蓬松感时倒梳发根使用，比尖尾梳温和、容易处理，对发质造成的伤害小，不易使头发打结。

It is a flat comb with a layer of soft bristle. It is used to backcomb the hair to make the hair fluffy. Compared with a pointed-tail comb, it is less aggressive and more manageable, causing less harm to the hair and reducing the chance of tangling.

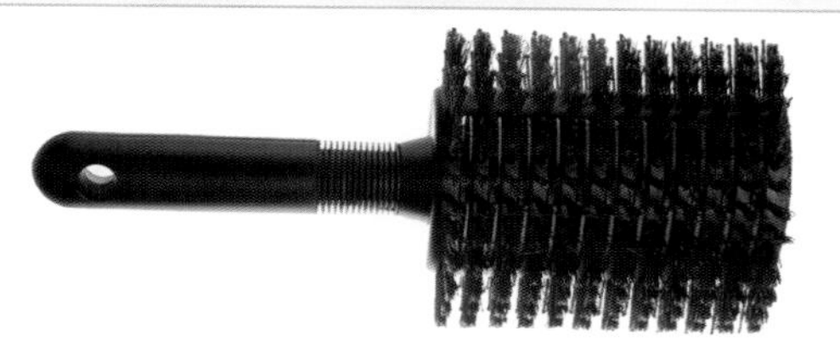

宽鬃毛梳 Broad bristle comb

用于梳理烫好的卷发，使卷发蓬松且更富弹性。

It is used to comb the perm to make the curls more fluffy and elastic.

S 形梳 S-shaped comb

用于梳顺毛糙的或打毛的头发。

It is used to comb the coarse and rough hair smooth.

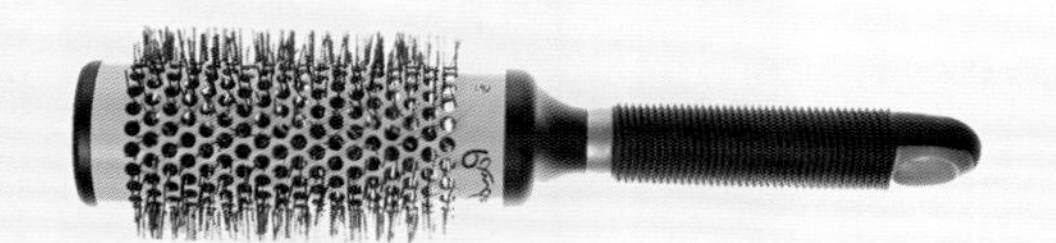

钢齿造型梳 Styling comb with steel teeth

与吹风机配合使用。这款梳子可以迅速平均分布热量，有助于在吹风的同时固定造型，是将头发吹顺或吹卷的造型利器，且不会因为长时间加热导致头发打结而损伤发质。

Working with a hair blower. This comb will quickly distribute heat evenly to set the hair shape. It is effective to help to blow the hair smooth or into curls without tangling or damaging the hair even after long-time heating.

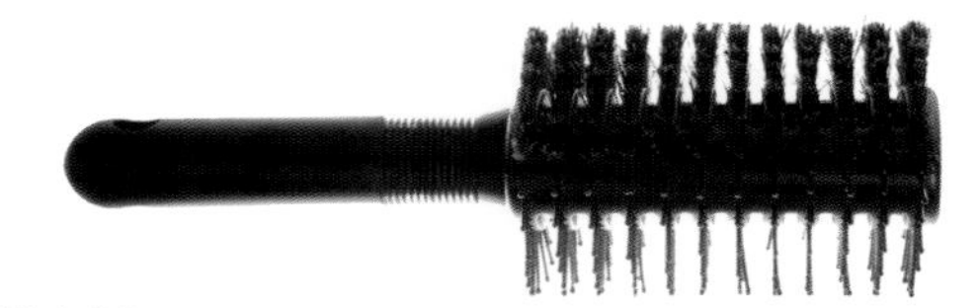

圆形鬃毛滚梳 Round bristle rolling comb

柔中带硬的刷毛材质，在梳理时除了能让秀发光滑平整，也能避免使发质受损。适合与吹风机搭配，可将头发吹出具有波浪线条的效果；也可将毛糙的卷发吹出顺滑的层次，使其呈现自然、蓬松、柔和的线条。

It is made of brush materials combined softness with hardness. It combs the hair smooth and neat without causing damage. It works well with a hair blower to blow the hair into waves or to make rough curls into smooth layers, presenting natural, fluffy, and soft lines.

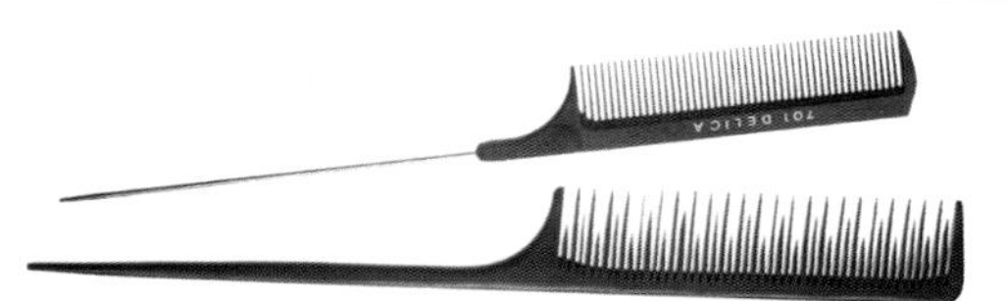

尖尾梳 Pointed-tail comb

是最基本的造型工具。尖尾部分可辅助头发划分区域，挑出造型中固定过紧的发丝，使其蓬松；前端可用于倒梳或梳顺发丝。是最常用的造型工具之一。

It is a basic styling tool. The pointed tail can help to divide the hair into sections and loosen up tightened hair. Its front end can be used to backcomb or smooth the hair. It is one of the most commonly used styling tools.

平板鬃毛梳 Paddle bristle comb

平板鬃毛梳的梳头又宽又平，适合梳理长发，尤其是长卷发。可将头发梳理得光滑顺泽，且不易变形。

The head of this comb is wide and flat, suitable for combing long hair, especially for long curly hair. It can make hair smooth and lustrous, and helps to maintain hair shape.

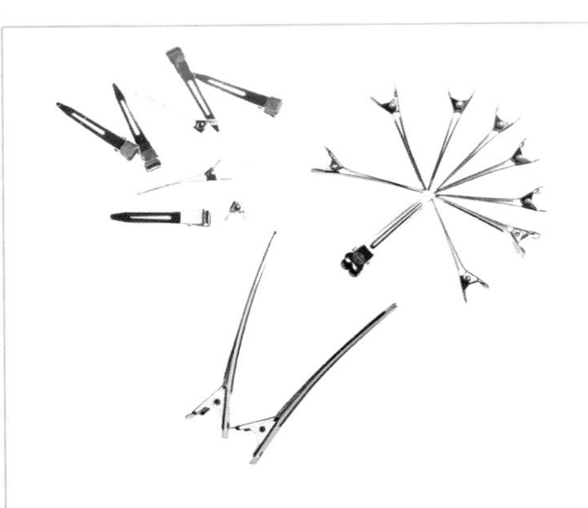

鸭嘴夹 Duckbilled hair clip

分带齿和不带齿两种，用来暂时固定分区的头发，可以协助造型。

It has two types: the toothed and the non-toothed. It is used for temporarily fastening the parted hair to facilitate the hair shaping.

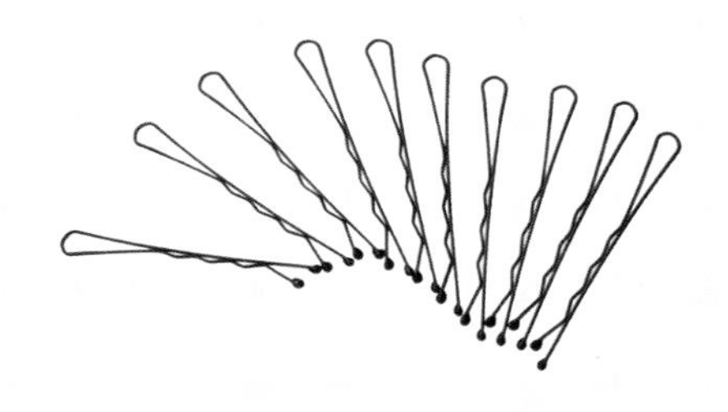

卡子 Pin

固定头发时使用。

It is used for fastening the hair.

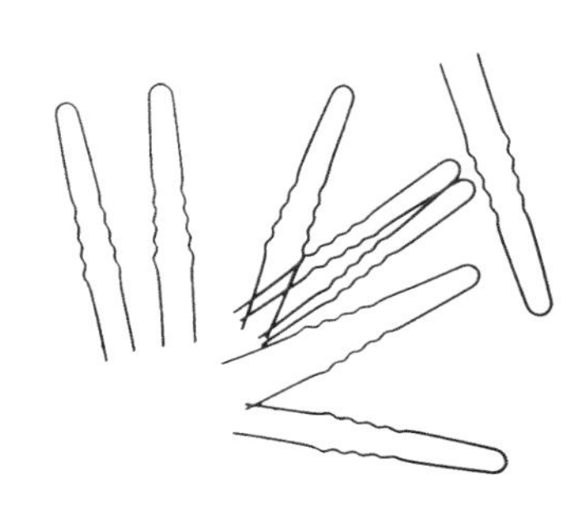

U 形卡 U-shaped pin

生活中不常用，主要用于固定一些造型较高的头发或连接较为蓬松的发丝。

It is uncommonly used in daily life. It is mainly used for fastening the hair with higher rise or connecting fluffy hair.

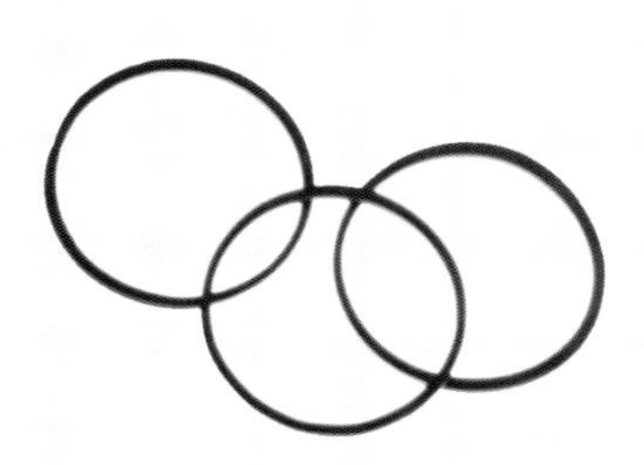

皮筋 Hair band

配合卡子固定头发，会将头发固定得很紧，不易扯伤头发。

It is used together with a steel clip to fasten the hair tightly, so that the hair won't be damaged from pulling.

无痕便利贴 No-trace hairpiece tape

用于固定刘海，防止刘海发丝变形。

It is used for securing bangs to prevent deformation.

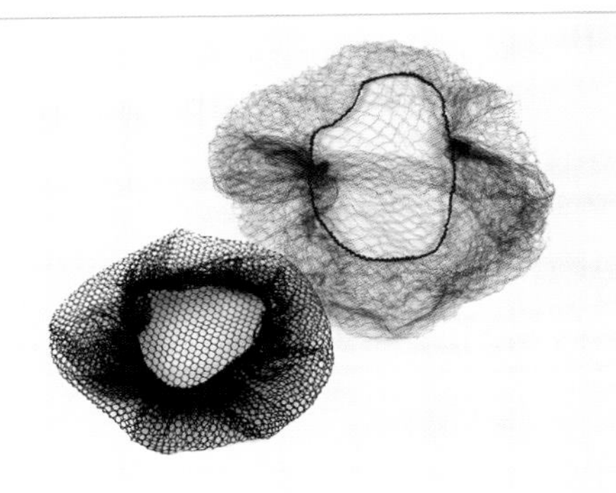

隐形发网 Invisible hairnet

用于包住头发，使头发表面干净，不易变形。一般用于设计发髻或佩戴假发之前，固定真发使用。

It is used to net the hair to make a neater and more stable surface. It is used generally for buns or for securing hair before putting on a wig.

造型发泥 Styling mud

真正的火山砂和蜡成分，可呈现自然的质感、高度的重塑性和自然的亚光效果，令造型看起来更加自然。编发前少量涂抹此产品，可有效避免毛糙感。

It is made of genuine volcanic sand and wax. It brings about natural elegance, high remoldability, and natural matte effect, making the hair style look more natural. Smearing some styling mud before braiding can prevent the hair from coarseness.

发胶 Hair spray

用于固定造型，快干、持久，自然不生硬，适合任何发质，在最后固定时使用。

It is used to set the hair into desirable shapes. It dries fast, lasts long, and presents natural softness, suitable for the final setting of hair of all textures.

造型蓬松粉 Styling powder

质地轻盈的粉末状造型产品，具有柔和的亚光效果，能吸附头发上多余的油脂，是用于蓬松发根的造型明星产品。

It is the powder with lightsome quality and soft matte effect. It can absorb excessive grease on the hair and make the hair fluffy at the roots. This is a star product for styling.

卷发器及效果图

THE CURLS TOOLS AND PERM EFFECT FIGURE

电烫棒 Curling wand

可以迅速改变发丝状态，平滑干枯毛糙的发丝，随意控制所需温度，令发卷更加生动，易于造型。

It can rapidly change hair condition, smooth dry and coarse hair, and freely adjust to the desirable temperature, making the curls more bouncy and vibrant for easier styling.

大号 Large size

32~38 号，使头发更富弹性，且时尚、自然，但持久性不强。

Size 32–38; creating fashionable and natural styles and making hair more bouncy; but the hair style won't maintain for long.

中号 Medium size

22~28 号，卷出的卷大小适中，给人端庄、优雅之感，适合盘发前使用。

Size 22–28; creating moderate-size curls, presenting a dignified, elegant style; being used before updos.

小号 Small size

9~13 号，小卷的烫发效果时尚感更强，适合发量较少的或需要增加头发蓬松度的造型。

Size 9–13; creating small ringlets with a more modern style, proper when the hair volume is moderate or more fluffiness is desirable.

直板夹 Flat hair clip

可将毛糙的头发夹直，也可用来卷发梢，效果较自然。

It can straighten coarse hair or create curls at the hair tips with a natural effect.

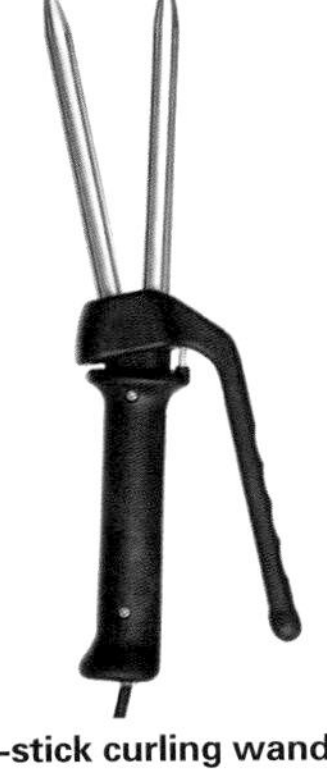

鸳鸯棒 Dual-stick curling wand

使烫发方式更加多变，发丝状态更加蓬松自然，一般用于时尚活泼的散发造型。

It is generally used for creating more fashionable and lively scattered hair styles. It can make the perm more dynamic, make the hair fluffier and more natural.

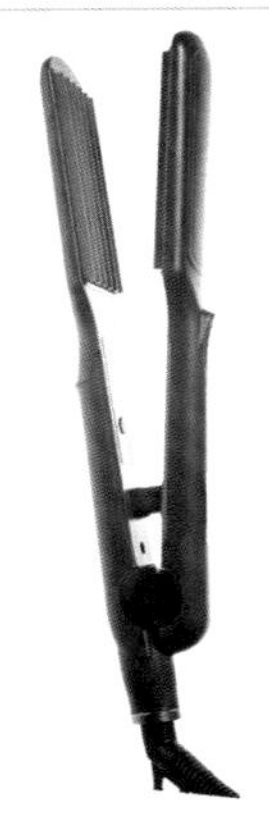

波纹夹 Wave-shaped hair clip

改变发丝纹路及方向，可增加发根蓬松度，使头发更蓬松，更易于盘发。

It can change the hair texture and direction, increasing the fluffiness of the hair at the roots for making updos easily.

恤发器 Hot air brush

可根据不同方式及方向取发片加热。因加热时间略长，可在妆前加热头发，使发丝变卷，更易于造型设计。

The use of this brush can be adjusted according to different styles and hair directions. Since it tends to take longer to heat up, it should be used before the making up to make the hair curly and more shapeable.

U 形棒 U-shaped wand

改变发丝状态，可为头发做出很卷的 S 形弯度，是日韩系流行的散发造型利器。

It is a styling tool for scattered hair, it's popular in Japan and South Korea. It can change the texture of the hair and roll the hair into large S-shaped curls.

魔术发卷 Magic hair curler

没有加热功能，可根据发型需要选择不同粗细的发卷。发卷表面的魔术贴直接接触头发，不要一次性卷太厚的头发。是制造自然发丝的必备工具。

It has no heating function. Curlers of different sizes can be selected according to the hairstyle. The velcro on the surface of the hair curler will directly contact with the hair. Don't roll too much hair on the curler for one time. It is a necessary tool to make natural hair style.

经典白纱造型设计解析

CLASSIC WEDDING DRESS HAIRSTYLE DESIGNS AND ANALYSIS

STYLING 造型步骤图 STEP GRAPH

STEP 1

将所有头发梳顺，固定于顶区偏左侧位置。

Comb all the hair smooth and secure the hair to the slight left side on the crown.

STEP 2

将所有头发均分为两份，将其中一份分层倒梳至发梢，以增加发量。

Divide all the hair into two parts and backcomb the hair on one side to trips to increase the hair volume.

STEP 3

将倒梳发片合拢，将表面梳理光滑，将发梢收拢。

Join the backcombed hair into a whole and comb the surface smooth; join the hair tips.

STEP 4

将梳理整齐的头发向后固定。

Secure the neatly combed hair backward.

STEP 5

将另一份头发同样倒梳，将表面梳理光滑。

Backcomb the hair on the other side in the same way and smooth up the surface.

STEP 6

以同样的方式固定头发，使两部分头发形成干净的蝴蝶结状。

Secure the hair in the same way and shape the hair of the two sections into a butterfly.

STEP 7

选择缎带花朵发卡，在中间头发分界线处佩戴。

Place clips of ribbony flower at the middle parting line.

整体造型以简洁干净的大结构盘发为主，蝴蝶结的造型设计打破了传统盘发的老气，时尚而俏皮。

This style highlights the neat and simple updo with a grand structure. The fashionable and nifty styling of a butterfly breaks the aged aura of a traditional updo.

STYLING 造型步骤图

STEP GRAPH

STEP 1
将头发分为前、后区，从后区开始向另一侧编发。
Divide the hair into the front and the back sections; braid the back hair towards the other side.

STEP 2
将发梢固定于另外一侧。
Secure the hair tips to the other side.

STEP 3
将编好的发辫固定于侧区。
Secure the braided hair on the side.

STEP 4
从刘海区开始，同样向侧区编发。
Starting from the bangs, braid the hair towards the side in the same way.

STEP 5
与侧区编好的发辫汇合固定。
Converge and secure together with the braided hair on the side.

STEP 6
将另一侧的刘海编发，固定于后区。
Braid the bangs of the other side and secure the hair on the back section.

STEP 7
用编好的假发发辫衔接填充，作为主要造型结构。
Use the braided wig to link and fill up the main styling structure.

STEP 8
选择小型珠花，点缀于发辫上。
Use small beaded flowers to adorn the braid.

整体造型时尚复古，编发以一侧为重点，点缀珠花，更显简洁。
This style is fashionable and retro. The braid is focused on one side and adorned with beaded flowers to give a neat hairdo.

STYLING 造型步骤图

STEP GRAPH

STEP 1

将所有头发梳顺，编发后固定于顶区，作为发型结构的基座。

Comb the hair smooth; braid and secure the hair on the crown as the structure base of the hairdo.

STEP 2

选取颜色与真发发色统一的卷曲假发，固定在顶区。

Choose a curly wig of the same hair color and fasten it on the crown.

STEP 3

将假发调整出所需的结构并固定。

Adjust the wig to the required structure and then fasten it.

STEP 4

调整假发的纹理及层次。

Adjust the wig lines and layers.

STEP 5

将调整好的层次用干胶固定，凸显假发的蓬松感及层次感。

Set the adjusted layers with spray to highlight the fluffiness and the layers of the wig.

STEP 6

在一侧点缀珠花发卡。

Adorn one side of the hair with a beaded flower hairpin.

整体造型以真假发结合的方式设计，时尚感十足，配合珠花点缀，时尚而大气。

This style combines the beauty of genuine hair and wig to present a trendy feeling. The beaded flowers lend more stylish and elegant aura.

STYLING 造型步骤图 STEP GRAPH

STEP 1

将所有头发梳顺，编发后固定于顶区，作为发型结构的基座。

Comb the hair smooth; braid and fix the hair on the crown as the structure base of the hairdo.

STEP 2

选取颜色与真发发色统一的卷曲假发，固定在顶区。

Choose a curly wig of the same color as the hair and fix it on the crown.

STEP 3

调整假发的纹理及层次，注意刘海区的随意性与真实性。

Adjust the wig lines and layers; pay special attention to arranging the bangs into a natural, real-life look.

STEP 4

选择缎带，以包发的形式固定。

Use a ribbon to fasten the hair in the form of wrapping.

STEP 5

调整发丝，固定好缎带。

Adjust the position of the hair line and fasten the ribbon.

STEP 6

再次调整发丝与缎带，在缎带固定处用钻饰发卡固定并点缀。

Adjust the hairline and the ribbon again and select a diamond hairpin to secure and adorn the ribbon.

整体造型以真假发结合为主，缎带头巾是时下流行的造型元素，复古中更显出时尚的味道。

This style combines the beauty of genuine hair and wig. Using the ribbon as a kerchief is very popular at present in order to make the hairdo look retro and stylish.

整体造型以对称的编发设计为主，简洁大方，钻饰珠花的点缀更能体现东方新娘的含蓄简约之美。
This style features symmetrical braiding. It is simple and gorgeous. When adorned with beaded flowers, it presents the beauty of modesty and simplicity of an oriental bride.

造型步骤图

STYLING STEP GRAPH

STEP 1
将头发中分，从一侧开始向后区编三股辫。
Part the hair in the middle and braid the hair on one side backward in a three-strand braid.

STEP 2
将编好的头发固定于后区。
Secure the braided hair on the back section.

STEP 3
取后区的头发，继续编发，与侧区连接固定。
Braid the hair on the back and secure it to the side section.

STEP 4
将另外一侧以同样的方式向后区编发。
Braid the hair on the other side backward in the same way.

STEP 5
编至后区时，从后区选取发片，以添加发辫的形式编发，注意发量及层次。
When the hair is braided to the back section, braid in some hair on the back, paying special attention to the hair volume and layers.

STEP 6
将后区剩余的头发编三股辫至发梢，注意使发辫尽量干净。
Make the rest hair on the back into a three-strand braid all the way to the hair tips; pay special attention to making the braid as neat as possible.

STEP 7
将编好的发辫向顶区固定，修饰外围弧度。
Fasten the braided hair on the crown and arrange the peripherals into proper shape.

STEP 8
利用假发辫再次修饰整体发型的外围轮廓。
Use wig braid to shape the peripherals further.

整体造型以简洁的三股辫设计为主，在额头处及顶区形成简洁的发髻，以钻饰珠花点缀于额头两侧，更能体现新娘的俏皮可爱。

This style uses neat three-strand braids to form a clean bun at the forehead and on the crown. When adorned with beaded flowers on both sides of the forehead, it fully presents the loveliness of the bride.

造型步骤图 STYLING STEP GRAPH

STEP 1

将所有头发梳顺，从刘海区开始以编发的形式向前提拉，编至发梢。

Comb the hair smooth; starting from the bangs, pull the hair forward and braid all the way to the hair tips.

STEP 2

将编好的发辫旋转固定于刘海处，在额头处形成蓬松的发髻。

Coil the braid and secure it at the bangs to form a fluffy bun at the forehead.

STEP 3

将后区头发向上固定，以三股辫的形式编至发梢。

Secure the back hair upward and make a three-strand braid all the way to the hair tips.

STEP 4

将后区编好的发辫与刘海衔接，固定于顶区位置，形成顶区高度。

Connect the braid on the back with the bangs and fix the braid on the crown to form a height.

STEP 5

将左、右侧区的头发向顶区收起，固定于顶区。

Collect and secure the hair of the right and left sides onto the crown.

STEP 6

将左、右侧区固定的头发合拢，以三股辫的形式编至发梢，修饰顶区的结构。

Combine the secured hair of the right and left sides and make a three-strand braid all the way to the hair tips to adorn the crown structure.

STEP 7

选择钻饰珠花，以不对称的形式点缀于刘海结构的两侧。

Use diamond beaded flowers to adorn the two sides of the bangs in an unsymmetrical way.

整体造型以简洁的包盘发结构为主，在结合处点缀花材，优雅、复古而灵动。

This style features a simple updo structure. The flower materials at the connections make the hairdo elegant, retro, and lively.

STYLING 造型步骤图

STEP GRAPH

STEP 1
将刘海区的头发倒梳，以拧包的形式固定。
Backcomb the hair of the bangs to secure the hair into a twisted bun.

STEP 2
将左侧的头发倒梳，将表面梳理光滑，以包发的形式固定于顶区。
Backcomb the hair on the left side to make the surface smooth and secure the hair into a bun on the crown.

STEP 3
以同样的方式将右侧区的头发固定于顶区，将后区的头发向上提拉并倒梳。
Secure the hair on the right side in the same way on the crown and backcomb the back hair while pulling the hair.

STEP 4
将后区的头发向上梳理，固定于顶区位置，使其形成一个饱满的结构。
Comb the back hair upward and secure the hair on the crown to form a full structure.

STEP 5
在发型结构连接处用花材点缀，并遮挡头发结构之间的空隙，取发梢修饰顶区。
Use flower materials to adorn the connection areas of the hair structure and hide the blank spaces in the hair structure. Take hair tips to decorate the crown.

整体造型简洁干净，花材以不对称的形式点缀，优雅而简洁。

This style is neat and simple. When adorned with flower materials asymmetrically, the hairdo looks elegant and neat.

STYLING 造型步骤图
STEP GRAPH

STEP 1
用直板夹处理发梢，使发梢自然内扣。将刘海以 2: 8 的比例分开，分层倒梳并固定，保证头发的蓬松度。
Treat the hair tips with a flat clip to make them naturally curled inward. Part the bangs to 2:8. Backcomb and secure the bangs by layers to ensure the fluffiness of the hair.

STEP 2
以同样的方式倒梳刘海发根，将表面梳理光滑，向一侧固定。
Backcomb the bangs at the roots in the same way. Comb the surface smooth and fix the hair to one side.

STEP 3
以同样的方式处理另外一侧的头发。
Treat the other side in the same way.

STEP 4
将发丝表面用手抓出层次，用发胶固定，增加其层次感。
Hand-arrange the hair surface to form layers and then set the hair surface with hair spray to increase the surface layers.

STEP 5
选择白色兰花，在头发发缝处点缀。
Adorn the parting lines with white orchids.

STEP 6
选择同样的花材，在另外一侧结构的连接处点缀，形成活泼的、不对称的形式。
Adorn the connection areas on the other side with the same flower materials to form lively asymmetry.

整体造型以单侧式的设计为重点，简洁大方。迎春花材隐约点缀于发丝间，更显新娘的妩媚气质。

This style features a lop-sided design. The structure is simple and elegant. The forsythia flowers showing in the hair further highlight the beauty of the bride.

造型步骤图 STYLING STEP GRAPH

STEP 1
将所有头发梳顺，烫外翻卷，注意侧区烫卷时要将层次尽量拉大。
Comb all the hair smooth and perm the hair into outward curls, pay special attention to making the spaces between the layers of the curls on the sides as large as possible.

STEP 2
将右侧的头发向左梳，以单侧式摆放，调整头发的层次。
Comb the right-side hair to the left and place it on one side; adjust the hair layers.

STEP 3
选择迎春花在耳侧固定。
Fasten forsythia flowers at the ear side.

STEP 4
调整发梢层次，使其覆盖于迎春花之上。
Adjust the layers of hair tips to cover the flowers.

STEP 5
将刘海区头发倒梳，将表面梳理光滑，以外翻的形式覆盖于迎春花上。
Backcomb the bangs, comb the hair surface smooth and arrange the hair into a backward wave to cover the flowers.

STEP 6
再次调整发梢层次，尽量使其蓬松自然。
Adjust the layers of the hair tips one more time to make the hairdo as fluffy and natural as possible.

整体造型以自然蓬松的手抓发为重点，简洁随意，体现出女性的端庄优雅之美。

This style features a natural, fluffy, hand-arranged hairdo. The structure is simple and casual, presenting elegance and dignified beauty.

造型步骤图 STYLING STEP GRAPH

STEP 1

将所有头发梳顺，用中号电卷棒烫卷。

Comb all the hair smooth and perm the hair with a medium-size curling wand.

STEP 2

将刘海区的头发倒梳，以手抓发的形式向上收拢，固定于顶区位置，发梢留出备用。

Backcomb the bangs and fold the hair upward by hand and secure the hair on the crown, reserving the hair tips for further treatment.

STEP 3

将后区的头发横向分发片，倒梳发根，使后区整体头发蓬松。

Divide the back hair crosswise into strands, backcomb the hair at the roots to make the back hair fluffy.

STEP 4

将后区的头发合拢，以手抓发和拧发的形式向顶区固定，发梢留出备用。

Combine the back hair, twist the hair by hand and secure the hair on the crown, reserving the hair tips for further treatment.

STEP 5

选择蝴蝶兰佩戴于顶区固定处，注意花材摆放的层次。

Place butterfly orchids at the anchoring point on the crown, pay special attention to the layers of the flowers.

STEP 6

将预留的发梢倒梳，处理出层次，衔接覆盖于花材之上。

Backcomb the reserved hair tips to create layers and cover them over the flowers.

整体造型以简洁的手抓发设计为主，发梢的蓬松层次使造型更为灵动而随意。

This style features a simple hand-made structure. The fluffy layers of the hair tips make the styling more lively and casual.

造型步骤图 STYLING STEP GRAPH

Step 1

将烫好卷的头发分成前、后区，从刘海区取发片，拧发并固定。将发梢留出，处理出自然的层次。

Divide the permed hair into front and back sections. Twist and secure strands of hair from the bangs; reserve the hair tips to create natural layers.

Step 2

以同样的方式从侧区取发片，拧发后固定于顶区。将发梢留出，处理出自然蓬松的层次，用干胶固定。

Twist the hair of the side likewise and secure the hair on the crown; make the reserved hair tips into natural and fluffy layers and set them with spray.

Step 3

从侧区开始以两股辫缠绕的方式向顶区固定。

Starting from the side, twine up the hair in a two-strand braid and fasten it on the crown.

Step 4

从后区横向分发片，倒梳发根，使其蓬松。

Divide the back hair crosswise into strands and backcomb the hair at the roots to achieve fluffiness.

Step 5

将倒梳好的头发收拢，固定于顶区，将发梢层次处理得自然蓬松。

Backcomb and secure the hair on the crown; arrange the layers of the hair tips to make them natural and fluffy.

Step 6

选择绣球花材点缀在鬓边一侧，将整体造型修饰得饱满圆润。

Select a hydrangea flower to adorn the temple area on one side; arrange and shape the overall structure to make it full and smooth.

整体造型自然随意，简洁大方，蓬松的卷发打破了传统盘发的呆板，粉嫩的花材更显俏皮可爱。
This style is natural and casual. The structure is simple and elegant. The fluffy curls break the stiffness of the traditional updo, and the pink flowers further increase the liveliness.

造型步骤图 STYLING STEP GRAPH

Step 1
将所有头发烫卷，分出前、后区，将后区头发竖向分发片倒梳，向顶区拧转并固定。
Perm all the hair to curls and divide the hair into front and back sections. Backcomb the back hair lengthwise strand by strand; twist and secure the hair on the crown.

Step 2
将后区固定后的头发的发梢层次调整蓬松，从一侧开始拧发，向顶区固定，用发梢修饰顶区。
Adjust the layers of the back hair tips to achieve fluffiness. Twist the hair from one side and fix the hair on the crown. Adorn the crown with hair tips.

Step 3
从刘海区横向分发片，倒梳发根，向顶区固定。
Divide the bangs crosswise into strands; backcomb the hair at the roots and secure the hair on the crown.

Step 4
将刘海区甩出的发梢的层次处理蓬松，衔接顶区与侧区。
Make the hair tips of the bangs into layers and achieve natural fluffiness; connect the hair with the crown and side sections.

Step 5
将另外一侧的头发分缕向顶区固定，选择粉色花材点缀于左右两侧。
Fix the hair of the other side strand by strand on the crown and adorn the right and left sides with pink flowers.

Step 6
将预留的发缕倒梳出层次，修饰整体发型轮廓。
Backcomb the reserved strands into layers; tidy up the overall structure.

整体造型以简洁的盘发为设计基础，随意而简洁，兰花的点缀更能体现造型的古典优雅风格。
This style features a neat updo. The structure is both simple and casual. The use of orchid better expresses the classical and elegant essence of the style.

造型步骤图 STYLING STEP GRAPH

STEP 1

将所有头发梳顺并烫卷，从刘海区开始向侧区以内扣的形式摆放。

Comb all the hair smooth and perm the hair into curls. Starting from the area of the bangs, create inward curls toward the sides.

STEP 2

从后区取发片，同样以内扣的形式摆放，衔接刘海区与侧区，注意尽量使层次丰富。

Take hair strands from the back section and place them likewise in inward curls and link them with the bangs at the side section; pay special attention to creating layers as rich as possible.

STEP 3

用顶后区的头发以拧发的方式修饰后枕骨处。

Twist the back hair on the crown to modify the back section.

STEP 4

将另外一侧的头发以同样的方式做层次卷，处理成发髻状并固定，注意尽量使发丝干净整齐。

Make layered curls on the other side in the same way and secure the hair into a bun, paying special attention to make the hair as neat as possible.

STEP 5

将剩余头发以空心卷的形式摆放并固定，修饰后区。

Create the remaining hair into hollow rolls and secure them to modify the back section.

STEP 6

选择兰花，在发型空缺处修饰。

Adorn the blank part of the structure with an orchid.

整体造型以松散的卷发与盘发相结合，随意而简洁，时尚而婉约。
This style features a combination of loose curls and updo. The structure is simple, casual, trendy, and modest.

造型步骤图 STYLING STEP GRAPH

STEP 1
将烫好卷的头发分为前、后区，取前区的头发，在头顶中间固定。
Divide the permed hair into front and back sections; take the front hair and fasten it in the middle.

STEP 2
将固定点的位置向下压，使其形成向上蓬起的刘海，发梢留出备用。
Press the hair downward at the fastening point to form upward fluffy bangs; reserve the hair tips for further use.

STEP 3
从侧区开始编松散的三股辫，向上收拢头发。
Starting from the side, make a loose three-strand braid and fold the hair upward.

STEP 4
将后区头发同样以松散的三股辫编发向顶区固定，发梢留出备用。
Braid the back hair in the same way and fix the hair on the crown; reserve the hair tips for further use.

STEP 5
以同样的手法处理另外一侧的头发，将所有头发的发梢汇聚于顶后区。
Treat the hair on the other side in the same way and fold all the hair tips to the back of the crown.

STEP 6
调整顶后区头发的层次，选择花材，点缀于鬓边衔接处。
Adjust the layers of the hair at the back of the crown; select flower materials to adorn the linking areas at the hairline at the temples.

STYLING 造型步骤图

STEP GRAPH

STEP 1

将头发烫卷，分成前、后两个区，从顶区开始横向分发片倒梳，甩出发梢，固定于顶区。

Perm the hair into curls and divide the hair into front and back sections; starting from the crown, divide the hair crosswise into strands to backcomb; leave the hair tips to be fixed on the crown.

STEP 2

在顶后区横向分发片倒梳，向顶区固定。

Divide the back hair on the crown into strands to backcomb, secure the hair on the crown.

STEP 3

以同样的方式倒梳，留出发梢，固定于顶区。

Backcomb the hair in the same way, leave the hair tips to be secured on the crown.

STEP 4

将刘海区中分，将侧区头发外翻后固定于顶区。

Part the bangs in the middle; turn the hair on both sides outward and secure it on the crown.

STEP 5

将侧区发梢与顶区汇合，保持发梢的卷，形成弧形。

Combine the hair tips of the sides with those on the crown, keep the curls of the hair tips, and create an arc–shaped structure.

STEP 6

刘海外侧以外翻的形式处理，衔接顶区及侧区。

Make one side of the bangs into an outward wave to reach the crown and the side sections.

STEP 7

以同样的方式处理另外一侧，保证造型的完整。

Treat the hair of the other side in the same way and ensure the fullness of the overall structure.

整体造型后区饱满，刘海区自然翻卷，顶区蓬松的卷发使整体造型更显妩媚，搭配白色兰花，简洁而优雅。

This style has a full back section. The bangs have natural curls. The fluffy curls on the crown will make the style more charming. When adorned with white orchid, the hairdo looks simple and elegant.

STYLING 造型步骤图
STEP GRAPH

STEP 1

将所有头发的发梢外翻烫卷，分成前、后区。
Perm the hair tips into curls and divide the hair into front and back sections.

STEP 2

从顶区开始横向分发片，倒梳至发根，使其蓬松。
Starting from the crown, divide the hair crosswise into strands and backcomb the hair all the way to the roots for fluffiness.

STEP 3

将顶区以包发的形式固定。
Secure the hair on the crown in a wrap.

STEP 4

从侧区横向取发片，以两股缠绕的方式固定在侧区，注意两股辫不要太紧，以增加自然蓬松度。
Take the hair of the sides crosswise and secure the hair on the side in a two-strand braid, paying special attention not to braiding too tight to achieve natural fluffiness.

STEP 5

将后区的头发向侧区固定，将发梢整理出随意卷曲的状态。
Secure the hair of the back section on the sides, and make the hair tips into casual curls.

STEP 6

将另外一侧的头发分发片倒梳发根，固定在顶区；发梢衔接侧区。
Backcomb the hair of the sides at the roots and secure it on the crown; connect the hair tips with the side sections.

STEP 7

选择白色兰花，填补发型的空缺位置。
Use white orchids to fill the blank of the structure.

整体造型卷盘结合，以一侧为重点，配以白色兰花，端正而随意。
This style combines curls with an updo. It focuses on one side. When adorned with white orchids, the hair looks elegant and casual.

SOFT WAN DYNAMIC CURLS WITH
ROMANTIC
PETALS
AS IF
BUTTERFLIES
FLYING
AROUND
THE HAIR
COMING
WITH
SWEET
AND
HAPPY JOY
和婉动的卷发
漫的花瓣仿佛飞上
边的蝴蝶
着甜美幸福的喜悦而来

星点珠花带来满心的喜悦 恰似水中精灵
Beaded flowers shine like stars and feel like watery spirits

整体造型简洁大方，利用外翻式刘海与侧区衔接，形成单侧式的发型结构，随意而灵动。

This style is both simple and elegant. A lop-sided shape is formed when the flip-out bangs connect with the side sections, creating a free and casual look.

STYLING 造型步骤图

STEP GRAPH

STEP 1

将头发烫卷，分出前、后区，取出刘海区，从侧区开始向后区固定。

Perm the hair into curls and divide the hair into front and back sections. Separate out the bangs. Starting from the sides, secure the hair on the back.

STEP 2

将后区的头发向一侧固定。

Secure the back hair to one side.

STEP 3

将侧区整理出层次。

Sort the hair on the sides into layers.

STEP 4

将刘海区的头发分发片倒梳，使整体刘海区蓬松。

Backcomb the bangs hair to make the entire bangs area fluffy.

STEP 5

将刘海表面梳理光滑，以外翻的形式与侧区连接。

Comb the surface of the bangs sleek and arrange the bangs in a flip-out wave to connect with the hair on the sides.

STEP 6

在刘海区与侧区的连接处佩戴饰品。

Place ornaments at the connection area between the bangs and the hair on the sides.

整体造型以蓬松的卷发处理，复古的刘海和饰品的选择增添了几分时尚与优雅。
This style features fluffy curls. The retro bangs and ornaments will add to the trendiness and elegance.

STYLING 造型步骤图

STEP GRAPH

STEP 1
选择大号电卷棒，将所有头发烫卷。
Use large-size curling wand to perm the hair.

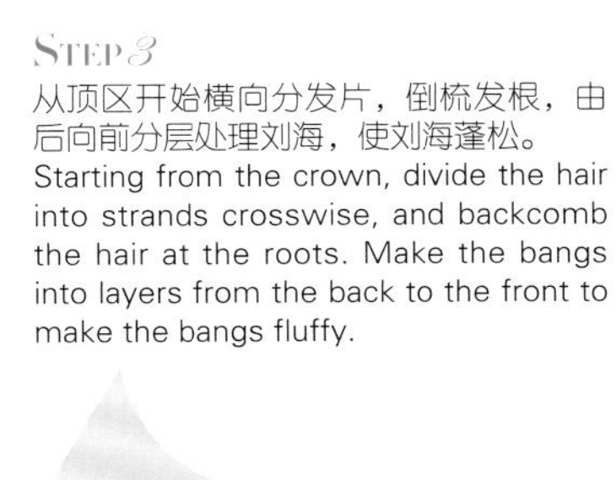

STEP 3
从顶区开始横向分发片，倒梳发根，由后向前分层处理刘海，使刘海蓬松。
Starting from the crown, divide the hair into strands crosswise, and backcomb the hair at the roots. Make the bangs into layers from the back to the front to make the bangs fluffy.

STEP 5
整理顶区与侧区的衔接处，不要有明显的分界线。
Arrange the hair at the connection area between the crown and the sides to eliminate any conspicuous parting line.

STEP 2
将左、右侧区的头发发根倒梳，使头发蓬松，层次分明。
Backcomb the hair of the two sides at the roots to make the hair fluffy with distinct layers.

STEP 4
将刘海向后梳理，与后区形成一体，用梳子尾端调整刘海的蓬松度及圆润度。
Comb the bangs backward to integrate with the back section; use the comb's end to adjust the fluffiness and fullness.

STEP 6
选择饰品修饰额头。
Select an ornament to adorn the forehead.

造型步骤图

STYLING STEP GRAPH

Step 1
将头发用恤发器烫卷。
Perm the hair with a hot air brush.

Step 2
将刘海中分，将后区横向分发片倒梳，使后区枕骨处的发型轮廓蓬松圆润。
Part the bangs in the middle; backcomb the back hair crosswise to achieve fullness.

Step 3
将烫好的头发用手分出层次。
Hand-arrange the permed hair into layers.

Step 4
将左、右刘海摆放成外翻卷，喷干胶固定。
Arrange the bangs at the left and right respectively into flip-out waves; set the waves with spray.

Step 5
在前、后区分界线处佩戴发带。
Place a ribbon at the parting line between the front and back sections.

Step 6
选择纱材，衔接在一侧外围轮廓处。
At one side, link the peripherals with mesh materials.

整体造型以自然的卷发设计，随意而灵动，活泼可爱的发卡式饰品更增添了造型的动感。
This style features casual curls. It is free and vibrant. The lovely hairpin design will increase the dynamic of the hairdo.

造型步骤图

STYLING STEP GRAPH

STEP 1

将所有头发烫卷，从一侧开始向上梳理，固定在顶区位置。

Perm the hair; comb the hair upward from one side and then secure the hair on the crown.

STEP 2

将刘海向右侧梳理，把后区头发分缕摆放出层次。

Put the bangs at the right side; and divide the back hair into layers.

STEP 3

调整层次的衔接处，利用发尾的随意卷衔接结构的不足。

Adjust the connecting areas of the layers and use the free-style curls at the ends to make up the insufficiency of the structure.

STEP 4

将刘海向右侧梳理干净，使其形成 S 形刘海。

Comb the bangs neat to the right side to the form S-shaped bangs.

STEP 5

从顶区位置取少量发片，倒梳至发根，在左侧也摆放出 S 形刘海。

Take a small amount of hair from the crown to backcomb all the way to the roots; make S-shaped bangs at the left side.

STEP 6

在整体结构与垂发衔接处佩戴珠花。

Use beaded flowers to adorn the overall structure and the connection areas of the hanging hair.

整体造型以复古的 S 形刘海为设计重点，随意的单侧式垂发更显女性的妩媚与性感。
This style emphasizes the retro S-shaped bangs. The casual, lop-sided, hanging-down hair increases the feminine charm and sexiness.

整体造型以干净的大结构盘发为主，真假发的结合更加简化了造型，皇冠以发卡的配搭方式打破传统，时尚而复古。
This style features a neat, big updo structure. The combination of the genuine hair and the wig further highlights the simplicity of the style. Using the tiara as a pin is a breakthrough over the tradition, presenting a trendy and retro aura.

STYLING 造型步骤图

STEP GRAPH

STEP 1

将刘海分发片倒梳，使整体刘海区蓬松。

Backcomb the bangs hair to make the entire bangs area fluffy.

STEP 2

将所有头发向一侧梳理光滑，用皮筋固定，注意固定在外耳廓以上的位置。

Comb all the hair smooth to one side and secure the hair above the ears with a hair band.

STEP 3

选择与真发发色、发丝状态接近的大刘海假发，固定于真发表面。

Choose a big-bangs style wig of which the color is close to the genuine hair and fasten it on the surface of the genuine hair.

STEP 4

将固定的真发和假发倒梳，增加头发的蓬松度，并将真假发衔接完整。

Backcomb the fastened genuine hair and the wig to increase the fluffiness of the hair and connect the genuine hair and the wig seamlessly.

STEP 5

将真假发合拢，将表面梳理光滑，并将发梢固定。

Combine the genuine hair and the wig, comb the surface smooth, and secure the hair tips.

STEP 6

将皇冠以发卡的佩戴方式佩戴在真假发固定的位置。

Pin a tiara at the anchor point of the genuine hair and the wig.

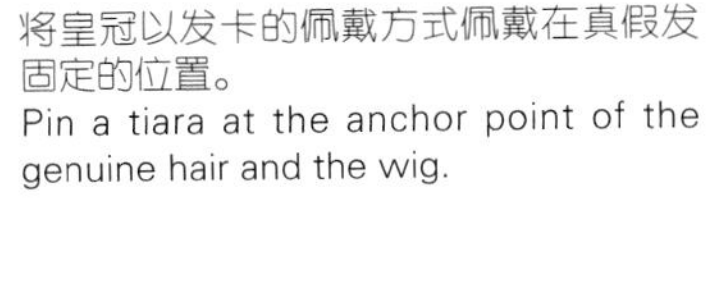

造型步骤图
STYLING STEP GRAPH

STEP 1
将头发烫卷，分为前、后区，将后区的头发用皮筋向顶区固定。
Perm the hair into curls, divide the hair into front and back sections, and secure the back hair with a hair band on the crown.

STEP 2
从刘海区开始分发片，倒梳发根，形成向上的刘海。
Divide the hair from the bangs into layers and backcomb the hair at the roots to form upward bangs.

STEP 3
将刘海发片向后压，使其形成S形。
Press layered hair of the bangs backwards to form an S shape.

STEP 4
将刘海发梢留出，向侧面调整层次。
Make the hair tips of the bangs into layers on the side.

STEP 5
将后区的头发倒梳，使其与刘海区衔接。
Backcomb the back hair to connect with the bangs.

STEP 6
在一侧选择绢花佩戴。
Adorn one side with silk flowers.

整体造型运用向上盘发的技巧，但无盘发痕迹，造型随意而大气，高耸向上的刘海设计能够拉长脸形。

This style uses the updo techniques without any trace of doing so. The styling is casual and gorgeous. The rising bangs will make the face look longer.

造型步骤图

STYLING STEP GRAPH

STEP 1

将头发烫卷，将刘海中分，从一侧开始倒梳，以外翻的形式固定。

Perm the hair to curls. Part the bangs in the middle. Starting from one side, backcomb and secure the bangs outward.

STEP 2

将另外一侧的刘海以同样的方式处理，注意头发的蓬松度及层次感。

Treat the bangs at the other side in the same way; pay special attention to the fluffiness and layers of the hair.

STEP 3

从一侧后区开始倒梳发根，向侧区固定，与侧区衔接。

Backcomb the back hair at the roots; secure the hair at the side to connect with the side section.

STEP 4

注意表面发丝的层次的摆放与走向，可留下随意的垂发。

Pay special attention to the placement and direction of the layers of the surface hair; some hair can be left hanging down freely.

STEP 5

在顶区位置佩戴绢花，注意不要佩戴在正中位置，偏侧佩戴更显随意。

Place a silk flower on the crown. In order to present a more free-looking style, the flower should not be placed exactly in the center.

STEP 6

选择头纱，佩戴在顶区位置。

Place a veil on the crown.

整体造型利用自然的大结构盘发设计，随意而简洁，自然垂下的发丝与头纱的呼应更显柔美。

This style features a grand structure of free-style updo. It is casual and simple. The naturally hanging-down hair with a veil may highlight the feminine beauty.

随意蓬松的卷发一直是婚纱造型的不二选择，童话般的发卷配合兰花的设计，更显女性的娇柔。
This style of casual and fluffy curls has always been a favorite choice for brides. The fairy-tale-like curls with orchids fully present the feminine beauty.

造型步骤图 STYLING

STEP GRAPH

STEP 1

将头发梳顺，中分，用大号电卷棒烫卷，注意后区应竖向烫发。

Comb the hair smooth, part the hair in the middle and then perm the hair with a large-size curling wand, paying special attention to perming the back hair lengthwise.

STEP 2

将顶区头发烫卷时尽可能靠近发根，增加顶区的蓬松度。

The permed curls on the crown should be as close to the hair roots as possible to increase the fluffiness.

STEP 3

用鬃毛梳将烫好的发卷处理得更加蓬松随意。

Arrange the permed curls with a bristle comb to increase the fluffiness and casualness.

STEP 4

从顶区开始倒梳发根，增加顶区的发量。

Starting from the hair on the crown, backcomb at the hair roots to increase the hair volume of the crown.

STEP 5

调整好整体发型的蓬松度，选择兰花佩戴于一侧。

Adjust the fluffiness of the overall hair style and place an orchid on one side.

STEP 6

注意花材摆放的位置及形状。

Pay special attention to the location and shape of the flower.

整体造型以干净的盘发为设计基础，简洁而复古，红玫瑰与紫藤的完美结合更能体现造型的古典韵味。

This style is based on a neat updo. The structure is both simple and retro. The red roses and wisteria will further bring out the classical and elegant essence.

造型步骤图 STYLING STEP GRAPH

STEP 1

将头发烫卷并梳顺，从后区开始取发片，以空心卷筒向上翻卷，固定于后侧区。

Perm the hair to curls and comb the hair smooth; starting from the back section, roll up strands of hair into hollow curls and secure them at the side of the back section.

STEP 2

选择紫藤，卷成发筒后固定于侧区。

Fasten a wisteria rolled in a curl on one side.

STEP 3

另外一侧以同样的方式挽发髻并固定于侧区。

Roll up the hair into a bun in the same way and secure the hair on the other side.

STEP 4

取少许发片，以空心卷衔接侧区的位置。

Make some hair into hollow curls to connect the side section.

STEP 5

从刘海区横向分发片，倒梳发根，将发梢处理成外翻卷筒，与侧区衔接。

Divide the bangs area crosswise into strands and backcomb the roots of the hair strand by strand. Treat the hair tips into outward curls to connect with the side section.

STEP 6

选择同款紫藤与小型红玫瑰，点缀整体发型。

Select a wisteria of same kind and small-size red roses to adorn the overall structure.

整体造型以光滑优雅的卷发为设计重点，单侧式的垂发设计打破了盘发的呆板，使造型更加优雅妩媚。

This style features smooth and elegant curls. The one-sided hanging hair breaks the dullness of a traditional updo and increases the elegance and charm.

造型步骤图 STYLING STEP GRAPH

STEP 1

将头发梳顺并烫卷，分出前、后区，从后区开始横向分发片，以外翻空心卷固定于侧区。

Comb the hair smooth and perm the hair into curls; divide the hair into front and back sections; starting from the back section, divide the hair crosswise into strands and create outward curls to be secured on one side.

STEP 2

另外一侧同样以空心卷设计，固定于后侧区。

On the other side, create the hair into hollow curls in the same way and secure them on the back side section.

STEP 3

从刘海区横向分发片，倒梳发根，将头发表面梳顺。

Divide the bangs crosswise into strands and backcomb the hair at the roots, then comb the hair surface sleek.

STEP 4

将梳顺的头发以S形手摆波纹向后区的卷筒衔接。

Hand-make the hair, which has been combed smooth, into an S-shaped wave to link with the curls on the back.

STEP 5

将后区预留的头发用宽鬃毛梳梳理顺滑，使其自然垂于一侧。

Comb the reserved back hair smooth with a broad bristle comb and let the hair hang down naturally at one side.

STEP 6

使另外一侧的刘海发梢与后区衔接，选择红色兰花点缀于侧区位置。

Connect the hair tips of the other side of the bangs with the back hair; use a red orchid to adorn the side section.

真假发结合的造型随意自然，衔接过渡真实，整体效果优雅而大方。

This style combines the genuine hair and wig. The structure is casual and natural, the transitional area looks life-like and the overall look is graceful.

STYLING 造型步骤图 STEP GRAPH

STEP 1

选择与真发发色接近的大波浪卷发，将发卷梳理自然后备用。

Select a big wave wig of the similar hair color and comb the curls to a natural look for future use.

STEP 2

将头发分为前、后区，将后区的头发固定成发髻。

Divide the hair into front and back sections and secure the back hair into a bun.

STEP 3

将假发固定在后区的固定点。

Fasten the wig at the anchor point in the back section.

STEP 4

将刘海横向分发片，倒梳发根，使其蓬松。

Divide the bangs crosswise into strands and backcomb the hair at the roots to make the hair fluffy.

STEP 5

将倒梳好的刘海区的头发收拢在一起，将表面梳理顺滑，以 S 形向后区假发处梳理衔接。

Fold the backcombed hair of the bangs together, comb the surface smooth, and create the hair into an S-shape to connect with the wig at the back.

STEP 6

将另外一侧同样横向分发片，倒梳发根，使其蓬松。

Likewise, divide the hair of the other side crosswise and backcomb at the roots to make the hair fluffy.

STEP 7

将梳理好的头发合拢在一起，用梳子在头发中部将发片向前推，后拉发梢，与后区假发衔接。

Fold the properly combed hair together, use a comb to push the hair forward from the middle part, and pull the hair tips backward to link with the wig in the back.

STEP 8

选择绢花花材佩戴。

Select silk flower materials to adorn the structure.

整体造型以优雅的发髻与皇冠相结合的手法处理，干净简洁，更能体现新娘的简约与端庄。
This style features a combination of elegant bun and tiara. The overall structure is neat and simple, highlighting the simplicity and gracefulness of the bride.

STYLING 造型步骤图 STEP GRAPH

STEP 1

将刘海发根倒梳，合并固定后一起向前推，使其形成高耸的刘海。

Backcomb the bangs at the roots, gather and fasten the hair together, and push the hair forward in order to form rising bangs.

STEP 2

选择圆形皇冠侧放在顶区位置，固定好皇冠。

Select a round tiara to place obliquely on the crown; secure the tiara.

STEP 3

将侧区的头发以外翻的形式与皇冠的底部边缘衔接并固定。

Flip out the side hair, connect and secure the hair with the bottom edge of the tiara.

STEP 4

将后侧区的头发同样以外翻包发的手法处理并固定，使其形成发髻。

Treat and secure the hair on the side of the back section in a likewise flip-out way to form a bun.

STEP 5

将另外一侧同样以外翻的形式与皇冠的底部边缘衔接并固定。

Secure the hair of the other side at the bottom edge of the tiara in the same manner.

STEP 6

将所有甩出的头发收拢，设计成下端式的发髻。

Gather all the left-out hair to create a bottom bun.

复古式发型的设计要点

无论是婚纱照拍摄还是结婚当天，古典的编发手法都适合东方新娘，皇冠与编盘设计更显新娘的古典气质。
The classical braid technique suits for the oriental bride in both the picture day and the wedding day. The tiara and braided bun will better present the classical grace of the bride.

STYLING 造型步骤图
STEP GRAPH

STEP 1
将所有头发梳顺，分为前、后区，将前区中分，从一侧向后区编发。
Comb all the hair smooth; divide the hair into front and back sections; part the front hair in the middle; starting from one side, braid the hair toward the back section.

STEP 2
另外一侧以同样的方式编发，在后区将两侧的头发合拢。
Braid the hair on the other side in the same manner, and gather the hair of both sides on the back section.

STEP 3
将两侧的发梢与后区的头发合拢，继续以三股辫编发。
After gathering hair tips of both sides and that of back section, continue to make a three-strand braid.

STEP 4
将编好的三股辫在后区以发髻的方式固定。
Secure the completed three-strand braid on the back section in the form of a bun.

STEP 5
选择皇冠佩戴在顶区位置，为突出时尚感，可适当向一侧摆放。
Place a tiara on the crown; in order to highlight a sense of fashion, place it on one side appropriately.

整体造型以编盘为主，后区的发髻设计与羽毛钻饰的配合使整体造型更具东方新娘韵味。
This style features braid and updo. The match of bun design of back section with feather and diamond ornament adds oriental aura to the bride.

STYLING 造型步骤图
STEP GRAPH

STEP 1
将所有头发梳理顺滑后，用波纹夹处理头发，注意可不处理头发表面。
After combing all the hair smooth, perm the hair with wave-shaped clips, pay special attention not to treating the surface of the hair.

STEP 2
从左侧刘海开始向右侧刘海及侧区编发。
Starting from the left side of the bangs, braid toward the right side of the bangs and the side section.

STEP 3
注意后面头发可适当蓬松，以便能形成饱满的发髻形状。
Pay attention to making the back hair properly fluffy to form a full bun.

STEP 4
将所有编好的发辫在后区固定成一个结构。
Secure all the braids into one whole structure on the back section.

STEP 5
在刘海与后侧区衔接处佩戴羽毛钻饰。
Adorn feathered ornaments on the connection area between the bangs and the back section.

STEP 6
注意饰品衔接在侧区与后区位置。
Pay attention that the ornaments are connected to the side section and the back section.

整体造型以简洁的盘发设计为主，蕾丝花边的帽饰设计更增添了新娘的优雅与古典气质。

This style features a simple updo. The lace headwear adds to the elegant and classical quality.

STYLING 造型步骤图

STEP GRAPH

STEP 1

将所有头发梳顺，用宽边蕾丝在额头处进行满包处理，将其固定。

Comb all the hair smooth; use wide lace to cover the forehead; secure the lace.

STEP 2

将头发一分为二，进行三股编发处理。

Divide the hair into two parts, and make three-strand braids.

STEP 3

将编好的发辫发梢收起并固定，形成后区的发髻。

Gather and secure the ends of the braid to form a bun on the back section.

STEP 4

将另外一侧以同样的方式固定。

Secure the other side in the same manner.

STEP 5

在一侧衔接处选择花材佩戴。

Select flowers to wear on the connection area of one side.

整体造型利用简洁的盘发为基础，结构简洁大气，配合白色网纱与龙胆，时尚而婉约。
This style bases its structure on a simple updo. It is simple and generous. When adorned with white mesh materials and gentian flowers, the style looks more fashionable and modest.

STYLING 造型步骤图
STEP GRAPH

STEP 1

将所有头发梳顺并烫卷，分出前、后区，从前区开始倒梳发根，向上梳理，使其形成蓬松的刘海区结构。

Comb all the hair smooth and perm the hair, divide the hair into front and back section; starting from the front section,backcomb the hair at the roots, and arrange the hair upwards to form a fluffy bangs structure.

STEP 2

在左右两侧鬓边选择白色龙胆花材佩戴，注意花材摆放的层次。

Select white gentian flowers to wear on the hairline of both sides,pay special attention to placing the flowers into layers.

STEP 3

从后区开始横向分发片倒梳，向刘海区固定，发梢留出备用。

Starting from the back section,divide the hair crosswise into strands to backcomb; secure the hair at the bangs section, and leave out the hair ends for further treatment.

STEP 4

将后区头发收拢，使其形成一个大的、饱满的结构，发梢部分甩出，作为外围轮廓修饰。

Gather the hair of back section to form a big and full structure, leave out the hair ends to decorate the peripheral lines.

STEP 5

选择大网纱与白色龙胆花材在脖颈处装饰。

Select a big mesh piece and white gentian flowers to adorn the neck area.

整体造型用蓬松自然的卷发作为设计重点，随意向上收拢，优雅而妩媚。

This structure features fluffy and casual curls. The hair is gathered upward casually to present elegance and loveliness.

造型步骤图 STYLING STEP GRAPH

STEP 1
将头发烫卷，取顶区位置的头发，倒梳发根处，使其在顶区形成一个饱满的结构。
Take the permed curls on the crown and backcomb the hair at the roots, so as to form a full structure on the crown.

STEP 2
用小号电卷棒烫卷，将所有发卷汇聚于顶区位置。
Create curls with a small-size curling wand and gather all the curls on the crown.

STEP 3
将后区头发倒梳，向顶区汇集并固定。
Backcomb one side of the back hair; gather and secure it on the crown.

STEP 4
另外一侧后区以同样的方式固定头发。
Secure the back hair on the other side in the same way.

STEP 5
调整好发型外围轮廓的弧度，为使造型显得更随意，可适当垂下一些发丝，修饰脖颈处。
Adjust the outline curve of the structure; in order to make the structure more casual, some hair may be allowed to hang down to adorn the neck.

STEP 6
选择花材在鬓边处点缀。
Select flower materials to adorn the hairlines at the temple areas.

珠花的完美运用

PERFECT APPLICATION OF BEADED FLOWERS

整体造型以随意的自然卷发为主，发辫在前额形成简单的点缀，时尚而婉约。

This style features casual, natural curly hair. The braid on the forehead from a simple adornment. The hairdo looks fashionable and modest.

STYLING 造型步骤图
STEP GRAPH

STEP 1
将头发梳顺，用大号电卷棒以内扣的形式卷烫。
Comb the hair smooth; perm the hair into inward curls with a large-size curling wand.

STEP 2
为使头发蓬松，用电夹板处理发根处，使头发更加蓬松。
Use perm browser to treat the hair at the roots to make the hair fluffier.

STEP 3
从一侧开始取发片，以三股辫编向另外一侧，提拉编发。
Starting from one side, take strands of hair to create a three-strand braid toward the other side while pulling the hair.

STEP 4
将编好的头发置于额头处，将发辫处理松散，将发辫发梢固定于刘海内侧。
Place the braided hair on the forehead, then make the braid looser, and last fasten the braid ends inside the bangs.

STEP 5
选择白色绣球花瓣点缀发辫及发缝衔接处，注意花材佩戴及外围轮廓的弧度。
Select white hydrangea petals to adorn the braid and hair parting lines; pay attention to the location of the flower materials and the curve of the peripheral line.

整体造型以卷发设计为基础，随意简洁，两侧的花材搭配更彰显出邻家女孩般的柔美。

This style is based on a curling hair design. The structure is casual and simple. The adornment of the flowers on both sides further presents the gentleness and beauty of a next-door girl.

造型步骤图 STYLING STEP GRAPH

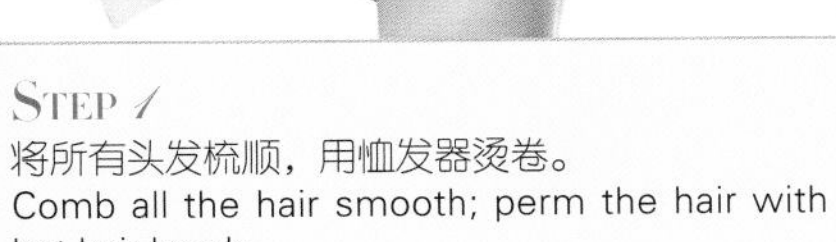

STEP 1

将所有头发梳顺，用恤发器烫卷。

Comb all the hair smooth; perm the hair with hot hair brush.

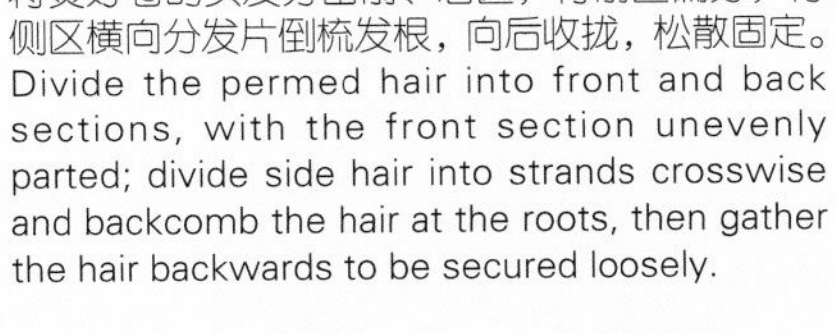

STEP 2

将烫好卷的头发分出前、后区，将前区偏分，将侧区横向分发片倒梳发根，向后收拢，松散固定。

Divide the permed hair into front and back sections, with the front section unevenly parted; divide side hair into strands crosswise and backcomb the hair at the roots, then gather the hair backwards to be secured loosely.

STEP 3

将另外一侧的头发以外翻的形式固定于侧区。

Flip up the hair of the other side and secure the hair on the side.

STEP 4

将顶区头发横向分发片倒梳，使其蓬松，覆盖于整个后区位置。

Divide the hair on the crown crosswise into strands, backcomb to make the hair fluffy and cover the whole back section.

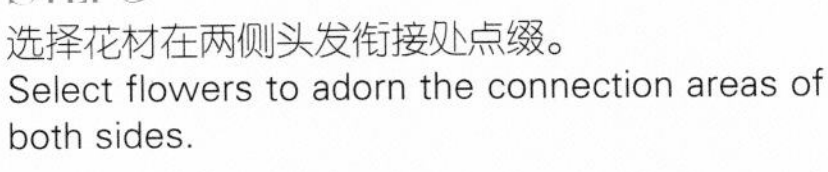

STEP 5

选择花材在两侧头发衔接处点缀。

Select flowers to adorn the connection areas of both sides.

STEP 6

将两侧垂发处理好，从后区取少量头发修饰侧区，将垂发层次处理得更加丰富。

After treating the hanging hair of both sides, take a small amount of hair from the back section to modify the side sections; treat hanging hair into rich layers.

整体造型利用随意的手抓发盘发技巧，配合花材使用，形成完整的造型，两侧丝丝缕缕的垂发更显自然随意。
This style uses the updo technique of casual hand-tossed hair. Adorned with flowers, the hairdo present a complete styling. The strands of hair falling naturally on both sides will increase the aura of naturalness and casualness.

造型步骤图 STYLING STEP GRAPH

STEP 1

将所有头发梳顺，选择大号电卷棒烫卷，将刘海区头发倒梳发根，向上固定成蓬松的结构。

Comb all the hair smooth; perm the hair with a large-size curling wand; backcomb the hair of the bangs section at the roots; secure the hair upward into a fluffy structure.

STEP 2

在刘海两侧及侧区留少量垂发，从侧区开始以手抓发的形式向顶区固定。

Leave a small amount of hair fall naturally on the side section and both sides of the bangs; starting from the side section, hand-toss and secure the hair to the crown.

STEP 3

在后区两侧分发片倒梳发根，将头发向顶区以手抓发的形式固定。

Divide the hair on both sides of the back section into strands; backcomb the hair at the roots; hand-toss and secure the hair to the crown.

STEP 4

固定后区发片，注意将发梢藏于发包内。

Secure the hair on the back; pay special attention to hiding the hair tips in the wrap.

STEP 5

选择蓝紫色花材，在发型外围轮廓上固定。

Select blue or violet flowers to fix around the peripheral line of the structure.

STEP 6

最后选择蓝色绣球花材，点缀及修饰外围轮廓。

Select a blue hydrangea flower to adorn the peripheral line of the structure.

整体造型以松散的手抓发为设计重点，采用不对称的花材佩戴方式，更能显示出女性的娇柔与妩媚。

This style features hand-tossed loose hairdo. It uses flowers in asymmetrical manner to further present the feminine gentleness and charm.

造型步骤图 STYLING STEP GRAPH

STEP 1

将所有头发梳顺，以恤发器烫卷，从刘海区开始横向分发片倒梳，将刘海分界线模糊掉。

Comb all the hair smooth; perm the hair into curls with a hot air brush; starting from the bangs section, divide the hair into strands crosswise to backcomb, so as to dim the boundary of the bangs.

STEP 2

将倒梳好的头发以手抓发的形式向后梳理出自然的层次。

Hand-toss the backcombed hair into natural layers.

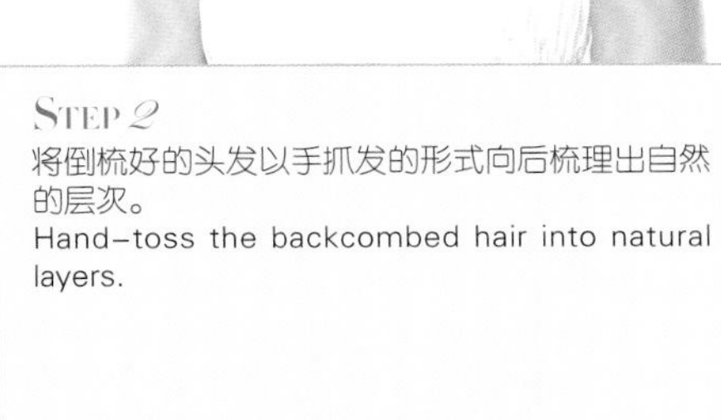

STEP 3

将两侧头发向后收拢，固定于后区。

Gather the hair of both sides backwards to secure on the back.

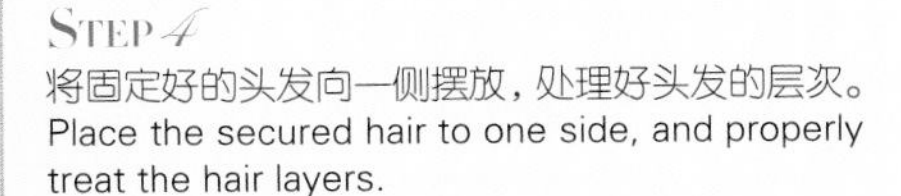

STEP 4

将固定好的头发向一侧摆放，处理好头发的层次。

Place the secured hair to one side, and properly treat the hair layers.

STEP 5

选择迎春花，在侧区垂发衔接处以放射状修饰。

Adorn the connection area of the side hanging hair with forsythia flowers in a radiant pattern.

STEP 6

另外一侧选择同款花材，以不对称式的手法修饰。

Select flowers of same type to decorate the other side in an asymmetrical manner.

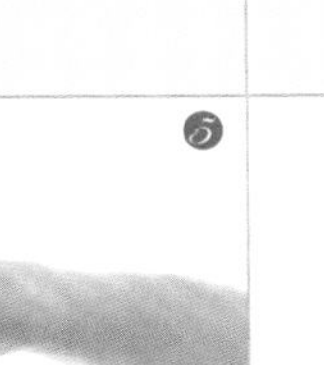

整体造型将简洁的盘发与蕾丝、花材完美结合，随意而优雅，时尚而简约。
This style perfectly combines a laced updo and flower materials. The structure is casual, elegant, fashionable, and simple.

造型步骤图 STYLING STEP GRAPH

STEP 1

将所有头发梳顺并烫卷，将刘海区及两侧区的头发以内扣的形式烫卷。

Comb all the hair smooth; perm the hair; curl the hair of the bangs and both sides inwards.

STEP 2

将烫好的头发向后梳理，选择白色蕾丝花边，在额头处以发带的形式修饰额头，左右两侧可适当预留少许发丝。

Comb the permed hair backward; select a white lace to decorate the forehead in the form of a hair band; a small amount of hair on the left and right sides may be reserved for later treatment.

STEP 3

将蕾丝花边固定好后，从后区开始取发片，横向倒梳后区头发。

After fastening the lace, starting from the back, take the back hair strand by strand crosswise to backcomb.

STEP 4

将倒梳好的头发向上包发，遮挡住蕾丝的固定点。

Wrap the backcombed hair upward, hiding the anchor point of the lace.

STEP 5

选择花材，在蕾丝与头发固定的衔接处点缀，并有效遮挡衔接位置。

Select flower materials to adorn the connection point of the lace and the hair and effectively cover the connection point.

STEP 6

将两侧预留的发丝处理好层次，使其自然地覆盖于花材表面及两侧位置。

Treat the reserved side hair into layers, and let the hair naturally cover the surface of the flowers and both sides.

整体造型以发卷打造外轮廓，让发型自然随意、时尚动感。钻饰的佩戴增加了高贵感。

This style uses curls to shape the outline of the structure to make the hairdo natural, casual, stylish, and dynamic. The use of diamond ornaments increases the aura of nobleness.

STYLING 造型步骤图
STEP GRAPH

STEP 1

将所有头发梳顺，选用大号卷棒，将头发全部烫卷至根部。

Comb the hair smooth; use a large-size curling wand to perm the whole hair all the way to the roots.

STEP 2

将刘海区的头发根部倒梳，让其有支撑力度。

Backcomb the roots of the bangs to make the area stiff.

STEP 3

将后区的头发分发束向上收拢，打造成随意的样式。

Divide the hair of the back into strands and gather the hair upward to create a casual look.

STEP 4

将左侧头发随意向上收拢，留出卷曲的发尾。

Gather up the hair on the left side casually, leaving out the curly hair ends.

STEP 5

将右侧余发全部向上收，留出卷曲的发尾，和上面的发卷衔接。

Gather all the remaining hair on the right side upward, leaving out the curly hair ends to reach the curls above.

STEP 6

调整发型外轮廓的蓬松度和随意感，两侧用尖尾梳挑出长短不一的发丝，增加动感．最后在额前佩戴钻饰。

Adjust the outline of the hairdo to present a look of fluffiness and casualness; pick out uneven hair pieces on both sides with a pointed tail comb to add a sense of movement; adorn the forehead with diamond ornaments.

整体造型利用简洁的盘发与 S 形卷的设计为重点，黑色鱼骨纱材与雏菊完美配合，更显出时尚神秘的气息。
This style features a simple undo and S-shaped rolls. The perfect match of black fishbone mesh and daisies presents an aura of fashion and mystery.

STYLING 造型步骤图 STEP GRAPH

STEP 1

将所有头发用恤发器烫卷，将刘海三七分，在后区横向分发片，倒梳发根，做内扣空心卷。注意尽量使表面光滑。

Curl all the hair with a hot air brush; part the bangs at the ratio of 3:7; take hair strands crosswise from the back section and backcomb the hair at the roots to make inward hollow curls; pay special attention to making the surface as sleek as possible.

STEP 2

顶区用同样的手法倒梳发根，使其蓬松，覆盖于后区头发表面。使后区结构蓬松饱满。

Backcomb the hair of the crown at the roots in the same way to make the hair fluffy and cover the surface of the back hair, so as to make the back structure fluffy and full.

STEP 3

将左侧区预留的头发以 S 形卷摆放固定，衔接后区发型空缺处，从右侧横向取发片，倒梳发根，衔接后区。

Place and fasten the hair reserved on the left side into a hand-made S-shaped roll, reaching the structural blank area of the back section; take strands of hair crosswise from the right side and backcomb the hair to connect the back section.

STEP 4

以同样的方式倒梳右侧区的头发发根，摆放出自然蓬松 S 形卷，衔接于侧区及后区，将刘海区头发倒梳，使其蓬松，将发梢自然外翻，与右侧区衔接，喷发胶固定。

Backcomb the hair at the roots on the right side in the same manner and arrange a natural and fluffy S-shaped roll to connect the side and the back area; backcomb the hair of bangs to make the hair fluffy; the hair tips should flip up naturally to reach the right side; set the hair with hairspray.

STEP 5

选择蓝紫色雏菊，在顶区佩戴，以增加整体造型外围轮廓的高度。

Select royal purple daisies to adorn the crown to add the height of the peripheral line of the whole structure.

STEP 6

选择黑色鱼骨纱网，修饰整个造型的外围轮廓。

Select a black fishbone mesh to decorate peripheral lines of the whole structure.

活泼俏皮的点缀式造型设计

LIVELY AND NIFTY DESIGN OF ORNAMENT AND SCULPT TYPE

造型步骤图 STYLING STEP GRAPH

STEP 1

将所有头发梳顺，选择大号电卷棒烫卷，并将头发分成前、后区，从侧区开始将发梢以伏贴外翻的手法与后区垂发衔接。

Comb all the hair smooth. Curl the hair with a large-size curling wand; part the hair into front and back sections; starting from the sides, flip up the hair tips to reach the hanging hair at the back.

STEP 2

将刘海横向分发片，倒梳发根，使其蓬松，发梢自然垂下，与侧区的头发衔接。

Divide the hair of the bangs crosswise into strands and backcomb the hair at the roots to make the hair fluffy; let the hair tips fall naturally to reach the side hair.

STEP 3

在左侧区预留出少许的头发，整体向一侧处理并固定，尽可能保证后区头发的蓬松度。

Reserve a small amount of hair on the left side; put the center of the whole structure to one side and fasten it; ensure the fluffiness of the back hair as possible.

STEP 4

将左侧预留的少许头发倒梳发根，使其自然垂下，保留发梢的层次。

Backcomb the small quantity of the reserved left hair at the roots; let the hair fall naturally and keep the layers of the hair tips.

STEP 5

将发梢层次覆盖于侧区，摆放好位置，喷干胶固定。

Cover the layered hair tips appropriately on the side section; set the shape with hairspray.

STEP 6

选择黄绿色花材及叶子，点缀于刘海发缝处。

Select fresh green flowers and leaves to adorn the parting line of the bangs.

整体造型以优雅的大波浪卷发为设计重点，蓬松自然的刘海与单侧式的处理效果更显新娘的妩媚动人。

This style features a grand structure of elegant large-wave rolls. The fluffy and natural bangs along with the one-sided structure further present the charm of the bride.

造型步骤图 STYLING STEP GRAPH

STEP 1

将所有头发用恤发器烫卷，在中间以圆形分出少许刘海，用皮筋固定于发中位置。

Curl all the hair with a hot air brush; separate a small amount of hair from the bangs in a round shape; secure the hair in the middle with a rubber band.

STEP 2

将固定好的刘海向一侧旋转，保留自然的刘海层次。

Rotate the secured bangs to one side, maintaining the natural bang layers.

STEP 3

从左侧区以松散的三股发辫编发至右侧，修饰额头位置，将发梢固定于刘海顶区。

Starting from the left side，create fluffy three-strand braid to the right side; trim the hair of the forehead properly; fasten the hair tips to the top section of the bangs.

STEP 4

两侧预留少许发丝，使其自然垂下，将后区剩余的头发拉至顶区，以松散的马尾固定于顶区，注意尽量不要用梳子，用手抓出自然的层次即可。

Let a small amount of hair on both sides to fall naturally, pull the remaining hair of the back section to the crown, and secure it on the crown in a fluffy ponytail; trying not to use a comb, hand-toss the hair into natural layers.

STEP 5

顶区以手抓发处理出头发的层次并固定，将多余的头发以松散的三股辫设计固定于一侧，发梢留出，衔接顶区。

Hand-toss the crown hair into layers to be fastened on the crown; braid the remaining hair into a fluffy three-strand braid design to be fastened on one side; leave out hair tips to connect with the crown.

STEP 6

选用带茎及叶子的白色芙蓉，点缀于顶区位置，用尖尾梳挑出少许发丝，使其垂于花材表面。

Select a white cotton rose with stem and leaves to adorn the crown; pick out a small amount of hair with pointed-tail comb to hang over the flower surface.

整体造型利用自然的手抓发为设计重点，随意而简洁，松散的发辫点缀其中，更显新娘邻家女孩般的魅力。

This style features natural, hand-tossing techniques. The structure is casual and simple. The loose braids are spotted in the hairdo to further present the charm of a next-door bride.

造型步骤图 STYLING STEP GRAPH

STEP 1

将头发梳顺，选用大号电卷棒将头发全部烫卷。

Comb the hair smooth; perm the hair with a large-size curling wand.

STEP 2

在中心位置分出顶区， 将顶区头发用皮筋固定。

Separate the crown section at the center of the hair; fasten the crown hair with a rubber band.

STEP 3

将顶区固定好的头发发梢倒梳蓬松，从皮筋固定位置下压，贴近头皮处固定。

Backcomb the tips of the fastened hair on the crown fluffy, and press the hair to the scalp from the position of the rubber band and secure the hair.

STEP 4

将后区头发向顶区收拢固定，留出发梢，调整好层次，将两侧头发向顶区方向倒梳并固定。

Gather and secure the back hair to the crown, leave out the hair tips, adjust the layers, and then backcomb the hair of both sides towards the crown and secure the hair.

STEP 5

选择皇冠式发卡，斜向戴在前额发际线处，修饰额头。

Place a tiara-type hairpin obliquely at the forehead hairline to adron the forehead.

STEP 6

调节好整体向上的发丝结构和层次，喷干胶再次固定。

Adjust the upward structural layers and set the hair with hairspray again.

整体造型以向上的大结构为主，发丝层次分明，用皇冠修饰额头，更显新娘的端庄与霸气。

The style features an upward grand structure with distinct layers. When adorned with a tiara, the hairdo will fully present an aura of dignity and dominance.

造型步骤图 STYLING STEP GRAPH

STEP 1

将所有头发烫卷，分出前、后区，前区以中分设计。

Perm all the hair; divide the hair into front and back sections; part the front section in the middle.

STEP 2

从后区开始横向分发片倒梳，使其饱满，设计出饱满的后区发髻。

Starting from the back section, divide the hair crosswise into strands, and backcomb to make the hair full; create a full bun in the back section.

STEP 3

从一侧开始横向分发片并倒梳发根，将发梢自然向后区收拢固定，衔接侧区与后区之间的空隙。

Starting from one side, divide the hair into strands crosswise to backcomb the hair at the roots; gather and secure the hair ends naturally to the back section to link the blank area between the side section and the back section.

STEP 4

将侧区最外层发丝向后区倒梳，覆盖于头发表面，使侧区头发自然垂下，调整好层次。右侧区手法同上。

Backcomb the outermost layer of the side hair toward the back section to cover the hair surface; let the side hair fall naturally; adjust the layer properly; treat the right side in the same manner.

STEP 5

另外一侧以同样的方式处理，注意侧区与后区的衔接。

Treat the other side in the same manner; pay special attention to the connection between the side section and the back section.

STEP 6

选择皇冠，佩戴在头顶中央的位置。

Place a tiara on the center position.

造型以对称的松散盘发为主，随意而简洁，中央的皇冠佩戴更显新娘的端庄。
This style features a fluffy, symmetrical updo. The structure is casual and neat. The tiara in the center position will give the bride an aura of dignified elegance.

造型步骤图 STYLING STEP GRAPH

STEP 1

将所有头发烫卷，从右侧区取适量发丝，分两股拧成两股辫。

Perm all the hair; take an appropriate amount of hair from the right side to separate into two strands and twist into a braid.

STEP 2

将两股辫一直向后区另一侧固定。

Secure the two-strand braid all the way on the other side of the back section.

STEP 3

将刘海 3:7 分，自然向侧区梳理，将发梢与侧区衔接，将左侧区的头发向后包发。

Part the bangs at 3:7, comb the hair to the side section naturally, connect the hair tips to the side section, and wrap the hair of the left side backward.

STEP 4

从垂发中横向取发片，向上以外翻形式处理，固定于侧区。

Take the hanging hair strand by strand crosswise, flip up the strands, and secure them on the side section.

STEP 5

将所有剩余头发继续以手打卷的形式摆放并固定。

Arrange and secure all the remaining hair in the form of hand-made curls.

STEP 6

选择小型皇冠，倾斜佩戴于一侧。

Place a small tiara obliquely on one side.

整体造型以干净简洁的盘发为设计重点，后区不对称式的发型设计与偏侧佩戴的皇冠相得益彰，顺滑的刘海更加凸显出了新娘的优雅与简约之美。

This style features a neat and simple updo. The asymmetrical design of the back section and the obliquely donned tiara complement each other well. The sleek bangs highlight the elegant and simple beauty of the bride.

STYLING STEP GRAPH
造型步骤图

STEP 1
将头顶区的头发向后做拧包结构，分别从两侧区向后区、向下提拉发片，做成卷筒。
Create a twisted bun structure backwards on the crown, and create hollow rolls on both sides while pulling hair strands downwards.

STEP 2
将后区卷筒摆放出层次。
Arrange the rolls on the back section into layers.

STEP 3
选择石竹梅，放入后区卷筒结构中。
Pin pink plum blossoms into the roll structure on the back section.

STEP 4
在两额角处留出少许发丝，增加线条感。
Leave out a small amount of hair pieces at the forehead hairline to add texture.

STEP 5
调整外轮廓的形状及花材的大小。
Adjust the peripheral lines and the shape, as well as the size of the flowers.

整体造型以后区的结构为重点，将花枝和卷筒融合在一起，营造出浪漫典雅的韵味。

This style focuses on the structure of the back section. It combines flowers and rolls to create a romantic and elegant aura.

C 造型对比图 COMPARISON PHOTOS OF DIFFERENT STYLES

STYLING STEP GRAPH

造型步骤图

STEP 1

把耳后区的头发纵向分发片，垂直于头皮提拉，向前做内扣卷。

Divide the hair at the back of ears lengthwise into strands; create inward curls while pulling the hair at a right angle with the scalp.

STEP 2

将刘海三七分，在前区表层斜向分发片，做外翻卷。

After parting the bangs at a ratio of 3:7, divide the surface hair of the front section obliquely to create outward rolls.

STEP 3

在前区内层纵向分发片，做内扣卷。

Divide the interior hair of the front section lengthwise into strands to create inward rolls.

STEP 4

从头顶区横向分发片，提拉 90° 向后烫卷。

Divide the hair on the crown crosswise into strands; perm the hair backward while pulling the hair at 90° .

STEP 5

选择花材和造型纱，摆放在前区发际线处。

Select flowers and styling mesh to adorn the front hairline.

整体造型运用不同区域的烫卷方法，发型层次丰富，整体感觉清新俏丽。

This style uses different curling techniques for different sections. The hairdo is rich in layers, presenting a fresh and pretty aura.

STYLING STEP GRAPH 造型步骤图

STEP 1

选择大号卷棒将头发烫卷，主要以外翻烫发为主。选择波纹板将发根烫蓬松。将顶区头发根部倒梳，内扣做出顶区包发结构。

Perm the hair with a large-size curling wand, creating mainly outward rolls. Make the hair at the roots fluffy by treating the hair with an electric wave-shaped clip. Backcomb the crown hair at the roots and curl the hair inward to make a wrap structure on the crown.

STEP 2

在顶区发包结构的基础上，将后区头发横向分片倒梳，将表面梳顺，向上做卷，与顶区衔接，形成饱满的后区结构，将左侧区头发斜向分成两片，将根部梳顺，做内扣S形手打卷，重点修饰脸型，并使其轮廓饱满。

Based on the wrap structure on the crown, divide the back hair at the roots crosswise into strands and backcomb the hair, comb the hair surface smooth, and make curls upward to reach the crown structure, so as to form a full back structure; divide the hair of the left side into two strands obliquely, and, respectively, trim and comb the hair at the roots to hand-make S-shaped inward rolls; pay specially attention to modify the facial shape to achieve fullness.

STEP 3

右侧区也将头发斜分区，做S形手打卷，发尾部分要保持自然的卷曲程度。整体根部要饱满，发尾线条要自然干净。将刘海根部倒梳，将表面梳顺，按照前区烫卷的样子整理出S形的弧度，将发尾做外翻处理。

Part the hair obliquely on the right side to hand-make S-shaped rolls, and the hair trips should be kept in natural curls. The roots of the hairdo need to be full and the lines of the hair tips natural and neat. Backcomb the bangs at the roots and comb the surface smooth; create an S-shaped curve in the same way metioned above; flip out the hair tips.

STEP 4

将虎头百合佩戴在后区与侧区衔接的位置，然后在顶区佩戴蓬松的层次型头纱。

Place a tiger lily on the connection area of the back and the side sections. Place a fluffy layered veil on the crown.

整体造型突出了新娘委婉而端庄的气质，卷盘结合的手法搭配素雅的珍珠颈饰能恰到好处地表达这一特点，在蓬松感和层次感极强的头饰的衬托下，造型显得尤为端庄大气。

The hairdo highlights the modest and dignified disposition of the bride. Combining curling and updo techniques and using simple but elegant pearl necklaces can help to make the point. With the fluffy and layered headdresses, the hairdo looks especially dignified and gorgeous.

STYLING STEP GRAPH

造型步骤图

STEP 1

选择大号卷棒将头发卷好，最好以内扣和外翻的手法相结合，重点烫中部和尾部。将发根用波纹板烫蓬松。将头发 3:7 分，用三股加辫向下编，设计出刘海的样式，一直编到侧区耳根部位。在编发的过程中主要用到的是头发的中间部分，不要编得太靠近根部。

Use a large-size curling wand to perm the hair; it's better to curl both inward and outward and focus on the middle and end parts of the hair. Perm the hair at the roots to achieve fluffiness with a wave-shaped clip. Part the hair at a ratio of 3:7. Create a bangs style by braiding a three-strand braid downward; braid all the way to the back section at the back of the ear. When braiding, mainly use the middle section of the hair; don't make the braid too close to the hair roots.

STEP 2

将左侧区的头发也通过三股加辫的形式来设计，先从根部开始，越往后越蓬松。

Use the hair of the left side to create a three-strand braid in the same way. The braid should start from the roots of the hair and get fluffier and fluffier when approaching the tips.

STEP 3

将顶区头发取出，同样用三股辫的手法编好并固定。其目的是方便佩戴头纱，并使垂发部分不那么厚。

Take the crown hair to create a three-strand braid in the same way; secure the braid. The purpose is to make it easier to attach the veil and make the hanging hair not so thick.

STEP 4

在右侧区和后区衔接处佩戴鲜花，主要起到衬托作用。在顶区佩戴层次型头纱。将发尾整理出透气灵动的效果。

Place fresh flowers at the connection area between the right side and back sections, so as to better present the hairdo. Place a layered veil on the crown. Arrange the hair tips to create a fluffy and dynamic effect.

编和卷的手法贯穿于整个发型，显得精致而随意。此款造型属于典型的鲜花造型。

This style uses braiding and curling techniques throughout the creation. The hairdo is both refined and casual. It is a typical method for fresh flower styling.

STYLING STEP GRAPH

造型步骤图

STEP 1

选择大号卷棒，将头发烫卷，将前、后区分开，取出顶区头发，扎低角度马尾。

Use a large-size curling wand to perm the hair; divide the front and back sections; take the crown hair to make a low angle ponytail.

STEP 2

将扎好的马尾分成四份，向四个方向做手打卷结构，用发卡固定。

Split the well-prepared ponytail in four strands to hand-make rolls to four directions; set the structure with clips.

STEP 3

将后区头发分发片倒梳，向上做手打卷，与顶区结构衔接。点缀钻卡即可。

Divide the back hair into strands, backcomb the hair and create curls to reach the crown structure. Adorn the hair with diamond clips.

STEP 4

将前区头发中分，将根部倒梳，将表面梳理顺滑，做出干净大气的刘海结构。将多余发尾做手打卷，与后区结构衔接。

Part the front hair in the middle; backcomb the hair at the roots and comb the hair smooth. Create a neat and grand bangs structure; hand-make rolls by using the remaining hair tips and connect them with the back structure.

优雅端庄的气质是此款新娘造型最大的特点。简单的配饰点缀其中，恰到好处。

An aura of elegance and dignity is the most prominent feature of this bridal hairdo. The simple adornments in the hair match the style perfectly.

浪漫蓬松的卷发搭配田园感觉的小碎花，突出了新娘的甜美风格。
This style matches romantic and fluffy curly hair with small flowers, so as to present the sweetness of the bride.

STYLING 造型步骤图

STEP GRAPH

STEP 1

选择大号电卷棒将头发烫卷。分出刘海区，将头发根部横向倒梳，向上做出饱满的结构。

Perm the hair with a large-size curling wand. Separate out the bangs section, and backcomb the hair at the roots crosswise. Make a full upward structure.

STEP 2

选择带枝叶的小碎花，固定在刘海区下卡处，整理枝叶线条。将两侧区的发丝向上下卡，与花材组合，将发尾缠绕在花间。

Select small flowers with stems and leaves to be fastened at the clip position of the bangs section; rearrange the lines of the stems and leaves. Clip the hair of both sides upward to match the flowers. Coil the hair tips in the middle of the flowers.

STEP 3

选择小碎花，佩戴在刘海两侧，衔接刘海与侧区。收起并整理侧区的发尾，使其与花材衔接。

Select small flowers to be placed on both sides of the bangs, so as to link the bangs with the sides. Collect the hair tips of the sides to be connected with the flowers.

STEP 4

将后区头发根部倒梳，梳顺表面，向上做包发结构，发尾甩出，与侧区衔接。注意枕骨处头发的饱满度。

Backcomb the hair roots at the back of the head. comb the surface smooth, make an upward wrap structure, flip out the hair tips to connect with the sides. Pay attention to the fullness of the back section near the neck.

STEP 5

选择花材，点缀在侧区与后区发尾造型之间。

Select flowers to adorn the areas between the sides and the hair tips of the back section.

高耸而凌乱的刘海是整个造型的亮点，通过水钻、亮粉来增加造型的时尚感。
This style features rising and scattered bangs. The use of sparkling sprinkles will increase the modern aura of the hairdo.

STYLING 造型步骤图

STEP GRAPH

STEP 1

将头发烫卷，根部用波纹板处理蓬松。以两眉峰为分界线分出刘海区，将根部倒梳，做出蓬松效果，将发尾整理出动感透气的螺旋式结构。

Perm the hair, and make the roots fluffy with wave-shaped clips; separate the bangs in line with eyebrows; backcomb the hair at the roots to create fluffy effect; arrange the hair tips into a dynamic and fluffy spiral structure.

STEP 2

将两侧区的头发梳顺，向斜上方提拉，拧好后下卡，发尾甩出。

Comb the hair of both sides smooth, twist the hair while pulling the hair obliquely upward, clip the hair, and flip out the hair tips.

STEP 3

将头发根部倒梳，做出蓬松效果。将发尾整理出动感透气的螺旋式结构。

Backcomb the hair at the roots to create fluffy effect. Arrange the hair tips into a dynamic and fluffy spiral structure.

STEP 4

将侧区发尾与刘海区结合。将后区头发根部倒梳后向上梳理，拧包，发尾甩出，跟顶区衔接。

Combine the side hair tips with the bangs section. Backcomb the back hair at the roots, comb and twist the hair upward into a wrap, and flip out the hair tips to connect with the crown.

STEP 5

选择钻饰发卡，固定在侧区与刘海区衔接处。

Select diamond hairpins to be fastened at the connection area between the side and the bangs.

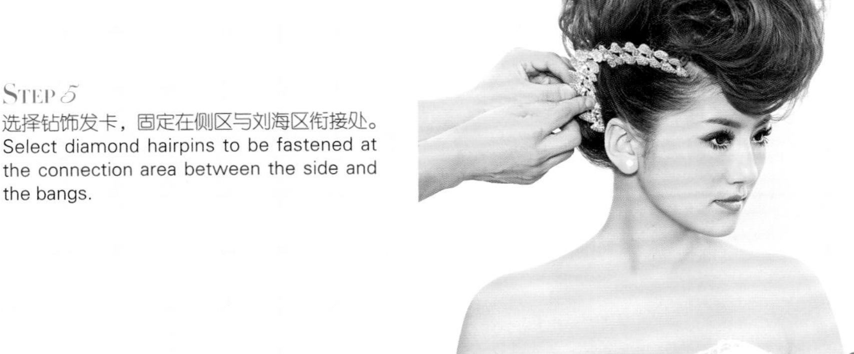

整体造型以简洁的盘发为设计重点，头纱与皇冠的完美搭配更显新娘的时尚与简约。
This style features a simple updo. The perfect match of the veil and the tiara adds to the fashion and modesty of the bride.

造型步骤图 STYLING STEP GRAPH

STEP 1
将所有头发烫卷，中分，将顶区头发处理蓬松，将所有头发向后梳理。
Perm all the hair; part the hair in the middle; after making the crown hair fluffy, comb all the hair backwards.

STEP 2
将后区头发外翻，挽出发髻的形状。
Flip the back hair outward to be tied into a bun.

STEP 3
将另外一侧后区的头发以同样的方式处理并固定。
Treat the hair of the other side of the back in the same manner; secure the hair.

STEP 4
将预留的头发向上以空心卷固定于后枕骨处，使其形成发髻。
Make the reserved hair upwards into hollow rolls to be secured at the lower back section to form a bun.

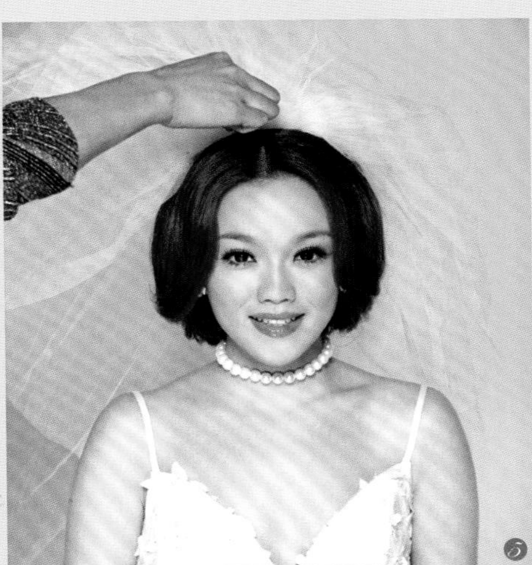

STEP 5
选择多层头纱，固定于顶区位置。
Fasten a multi-layer veil on the crown.

STEP 6
选择小型皇冠，佩戴于头发与头纱的衔接处。
Place a small tiara at the connection between the hair and the veil.

优雅大气的“赫本头 ”一直是新娘造型的不二之选，简洁的结构与皇冠的完美结合更显新娘女王般的气质。
Elegant and bounteous "Hepburn Style" has always been a favorite for brides; Moreover, the perfect match of the simple structure with the tiara presents the bride an aura of a queen.

造型步骤图 STYLING STEP GRAPH

STEP 1

将头发梳顺，固定于顶区位置，注意一定要牢固。
Comb the hair smooth and fix the hair on the crown firmly.

STEP 2

将固定在顶区的头发分成两部分，横向倒梳发片，注意发片的蓬松度。
Divide the hair secured on the crown into two parts, and backcomb the strands crosswise; pay special attention to the fluffiness of the strands.

STEP 3

将倒梳好的发片向后包发，轻轻将发片以扇面形展开并固定。
Wrap the backcombed strands backward into a bun; gently spread and set the strands into the shape of a fan.

STEP 4

将另一部分头发向前提拉，倒梳发根。
Backcomb the other part of hair at the roots while pulling the hair forward.

STEP 5

将倒梳好的头发表面梳理光滑，调整好蓬松度，固定发梢。
Comb the surface of the backcombed hair sleek, adjust the fluffiness, and secure the hair tips.

STEP 6

选择皇冠，固定在顶区位置。
Place a tiara on the crown of the head.

整体造型将干净的空心卷与网纱完美配合，造型简洁，体现出新娘的时尚简约之美。
This style features perfect match of simple hollow rolls and mesh. The structure is simple and neat, adding an aura of fashionable and modest beauty to the bride.

STYLING 造型步骤图

STEP GRAPH

STEP 1
将头发梳顺，固定于顶区位置。
Comb the hair smooth, secure the hair on the crown.

STEP 2
将固定好的头发横向倒梳，处理蓬松。
Backcomb the secured hair crosswise to make the hair fluffy.

STEP 3
将倒梳的头发表面梳理光滑，以空心卷的形式向一侧固定。
Comb the surface of the backcombed hair sleek, and secure the hair to one side in the form of hollow rolls.

STEP 4
选择白色大网纱，固定在头发空缺的一侧。
Fasten a white large-hole mesh on the blank side of the structure.

STEP 5
注意网纱摆放的位置及层次。
Pay special attention to the position and layer of the mesh.

垂纱的不同佩戴方式
Different styles of wearing veil

花材的组合与元素搭配

Combination of flower materials and arrangement of different elements

优雅曼妙的垂纱设计
An elegant and
graceful design of bridal veil

多变的时尚新娘

A CHANGEFUL FASHIONABLE BRIDE

清新优雅的公主造型
浓浓的贵族气质
Fresh and elegant
styles to make the bride like a princess with dense noble aura

经典晚礼造型设计解析

CLASSIC EVENING DERSS HAIRSTYLE DESIGNS AND ANALYSIS

时尚 唯美 简约 优雅

FASHIONABLE AESTHETIC SIMPLE ELEGANT

整体造型干净自然，发丝蓬松，垂发为整体造型增添了年轻的气息。
This style is neat and natural. The hair is fluffy. The hanging-down hair adds a youthful aura.

STYLING 造型步骤图

STEP GRAPH

Step 1

用电卷棒将所有头发烫卷，将刘海区 2:8 分开。

Use a curling wand to perm the hair; part the bangs to 2:8.

Step 2

将刘海区头发的发根倒梳。

Backcomb the bangs at the hair roots.

Step 3

将刘海表面梳理光滑整齐，向后区及侧区固定。

Comb the bangs surface smooth and neat and secure the hair to the back and the side sections.

Step 4

将另外一侧头发倒梳并固定，将表面处理光滑干净。

Backcomb and secure the hair on the other side and comb the surface smooth and neat.

Step 5

将外侧头发分缕覆盖在左侧光滑的发丝表面，形成层次。

Place the outside hair strand by strand on the surface of the smooth hair on the left side to form layers.

Step 6

选择羽毛花材，固定在发缝分界线处，将垂下的头发处理光滑。

Use feather materials to adorn the parting line and then comb the hanging-down hair smooth.

整体造型以大结构包发为主，外翻的刘海设计与蕾丝发带的完美配合更加凸显出女性的古典美。
This style features a grand structure of hair wrap. The perfect combination of the outward bangs and lace hair bands further presents the classical feminine beauty.

STYLING 造型步骤图 STEP GRAPH

STEP 1

将头发分为前、后区，从后区开始，横向分发片倒梳。

Divide the hair into front and back sections; start from the back section, divide the hair crosswise into strands and backcomb.

STEP 2

将所有后区头发合拢，在后区以包发设计。

Put all the back hair together and secure the hair on the back into a wrap.

STEP 3

从左、右侧区横向分发片，倒梳发根，使其蓬松。

Divide the hair crosswise strands from the right and left sides; backcomb the hair at the roots for fluffiness.

STEP 4

将包发表面梳理光滑，注意整体包发的圆润性。

Comb the surface of the hair of the wrap smooth; pay special attention to make the overall fullness of the wrap.

STEP 5

将覆盖在表面的发梢收拢、固定。尽可能使其随意自然，有透气感。

Collect and secure the hair tips covering the surface and arrange the hair tips as natural and fluffy as possible.

STEP 6

将刘海发根倒梳，增加刘海蓬松度。

Backcomb the bangs at the roots to increase the flulfiness.

STEP 7

选择蕾丝发带，在前、后区分界线处点缀。

Select a lace hair band to adorn at the parting line between the front and back sections.

STEP 8

将左、右刘海设计为外翻样式，与蕾丝发带衔接。

Part the bangs into the left and right sections and create them into outward waves to be fastened with a lace hair band.

整体造型层次丰富，配以粉色花材，将新娘凸显得俏皮而活泼。
This style creates rich layers. The pink flowers add to the lightness and liveliness of the bride.

STYLING 造型步骤图

STEP GRAPH

STEP 1

用电卷棒将所有头发烫卷，将左、右侧区的头发以外翻烫发处理。

Use a curling wand to perm the hair; pay special attention to creat flip-up waves out of both the left and right sides of the hair.

STEP 3

将左侧区的头发处理成蓬松的结构，使表面干净整齐。

Make the left-side hair into a fluffy structure with clean and neat surface.

STEP 5

选择粉色花材，分别佩戴在左、右侧区。

Select pink flowers to adorn the right and left sides.

STEP 2

将整体头发以 2:8 分区，将左侧前区的头发预留出来，竖向分发片倒梳。

Part the hair to 2:8, reserve the front hair at the left side for further treatment, divide the hair lengthwise into strands, and backcomb the hair.

STEP 4

将右侧区的头发做外翻处理，向上固定，预留出一缕垂发。

Create an outward curl out of the right-side hair and secure the hair upward, leaving a strand of hair hanging down.

STEP 6

将预留出的发丝覆盖在花材表面，使其形成自然的层次。

Use the reserved hair to cover the flowers to form natural layers.

整体造型利用真假发结合的方式处理，韩式的盘发技巧融入其中，更显新娘的端庄与柔美。
This style combines genuine hair and wig. It uses Korean-style updo skills to highlight the elegance and gentleness of the bride.

STYLING 造型步骤图
STEP GRAPH

STEP 1

将头发烫卷，分为前、后区，将后区头发固定，选择与真发颜色及发丝状态一致的假发，固定于后区位置。

Perm the hair to curls and divide the hair into front and back sections; select a wig with the same color and condition as the genuine hair to be fastened on the back section.

STEP 2

将假发梳理顺滑，调整好发丝状态，从侧区横向分发片倒梳发根，使其蓬松。

Comb the wig smooth and adjust the hair properly; start from the side, divide the hair crosswise into strands and backcomb the hair at the roots to make the hair fluffy.

STEP 3

将倒梳后的头发表面梳理光滑，将后区的假发做包发处理。

Comb the surface of the backcombed hair smooth and wrap the wig of the back section.

STEP 4

以同样的方法处理另外一侧的头发，真假发的衔接应尽可能自然。

Treat the hair of the other side in the same way, pay special attention to make the linking areas of the genuine hair and the wig as natural as possible.

STEP 5

将侧区预留的头发倒梳，修饰侧区与后区的衔接处。

Backcomb the reserved side hair; properly adjust the connection area between the side and the back.

STEP 6

将刘海区的头发偏分，将表面梳顺，将发梢以自然的外翻卷衔接于侧区位置并固定。

Part the bangs into two uneven sections, comb the surface smooth, and link the hair tips in natural outward curls with the side section.

STEP 7

将另外一侧的刘海以同样的手法衔接侧区。

Link the bangs of the other side in the same way with the side section.

STEP 8

选择紫色绢花佩戴于一侧。

Place purple silk flower on one side.

整体造型轻盈自然，利用发梢的层次增加造型的随意性，时尚而自由。

This style is lightness and natural. It uses the casual layers created by the hair tips to present a trendy and free aura.

STYLING 造型步骤图

STEP GRAPH

STEP 1

将烫好卷的头发分为前、后区，将后区的头发用皮筋固定。

Divide the permed hair into front and back sections; secure the back hair with a hair band.

STEP 2

将固定好的后区头发倒梳，按烫卷方向处理出层次。

Backcomb the secured back hair and create layers in the same direction as the permed curls.

STEP 3

将处理好层次的头发摆放在适合的位置，可用发卡再次向顶区固定。

Place the well-treated layers at the proper position; secure the hair to the crown with a hairpin if needed.

STEP 4

从侧区开始以包发形式向顶区固定，注意将发梢部分甩出，与顶区衔接。

Start from the side，secure the hair on the crown in the form of a wrap and leave out the hair tips to be connected with the crown.

STEP 5

将另一侧发根倒梳，以同样的方式将头发向顶区固定，发梢与顶区衔接。

Backcomb upward the hair of the other side and secure the hair on the crown in the same way; connect the hair tips with the crown.

STEP 6

侧区以包发的手法设计。

Arrange the side hair into a wrap.

STEP 7

左右两侧分别垂下一些发丝，增加造型的随意性。

Let some hair hang down naturally on both sides to increase the casualness.

STEP 8

选择与服装同色系的纱帽，佩戴于侧区。

Select a mesh hat of the same color as the dress and place it on the side.

整体造型以蓬松的卷发为设计重点，造型随意自然，更显新娘的优雅气质。
This style focuses on fluffy curls. It is casual and natural, presenting the bride's grace.

STYLING 造型步骤图 STEP GRAPH

STEP 1

将所有头发梳顺，选择大号电卷棒烫卷，将刘海偏分，从右侧开始横向分发片倒梳，使头发蓬松。

Comb the hair smooth; perm the hair with a large-size curling wand; part the bangs into two uneven sections; start from the right side, divide the hair crosswise into strands and backcomb the hair fluffy.

STEP 2

将倒梳好的头发表面梳理光滑，侧区预留出少许发丝。

Comb the backcombed hair to make the surface sleek, reserve some hair at the side for further treatment.

STEP 3

用梳子尾端向后压住头发，下发卡固定，使其形成蓬松的结构。

Use the end of the comb to press the hair backward and fasten the hair with a hairpin to form a fluffy structure.

STEP 4

将预留好的垂发层次处理均匀。

Treat the reserved hanging-down hair into balanced layers.

STEP 5

将刘海另外一侧头发梳顺，横向取发片倒梳，使刘海蓬松。

Comb the other side of the bangs smooth; divide the hair crosswise into strands and backcomb and make the bangs fluffy.

STEP 6

将倒梳好的发片合拢并梳顺。

Put the backcombed hair together and comb the hair smooth.

STEP 7

将发梢以外翻形式固定，整理好头发层次。

Set the hair tips into an outward wave and sort out the hair layers.

STEP 8

选择宝蓝色绢花佩戴于刘海一侧。

Select a sapphire silk flower to be worn at one side of the bangs.

气质型优雅晚礼
简约随意风格
更加彰显新娘的温婉柔美
Elegant evening wear with a
simple and casual style display the bride's
gentleness and beauty.

优雅的手打卷发搭配浅绿色的花冠，如早春庭院中漫步的公主，清新而不失典雅，高贵却不失稚嫩可爱。这款造型重点在于手打卷发的流畅弧度及发卷的光泽感，而蓬松的顶区轮廓削弱了发型的年龄感。

Elegant hand-made curls going with light green flower crown, pure and fresh, elegant and noble, just like the hair of a princess strolling in a courtyard in early spring. This style features smooth curves created by hand-made curls as well as the luster of the curls. The fluffy crown relieves the sense of agedness of the style.

STYLING 造型步骤图 STEP GRAPH

STEP 1

将发丝用恤发器烫卷，分为前、后区，将后区头发固定，将发尾倒梳，做成蓬松状。使发型侧面更加饱满。

Curl the hair with a hot air brush and divide the hair into front and back sections; secure the back sections and backcomb the hair tips fluffy to increase the fullness of the sides.

STEP 2

将侧区发丝倒梳，将表面梳理光滑，先将发根向上梳理，再翻卷发尾，使其呈 S 状。喷发胶固定。

Backcomb the hair of sides and comb the surface smooth, comb the hair roots upward first, turn and roll the hair tips into an S-shape. Set the hair with spray.

STEP 3

将另一侧的发丝分出一部分，向上提拉并倒梳发根。要使发卷弧度更大、更流畅，需向发卷相反的方向提拉。

Separate the crown hair of the other side, backcomb the hair roots while pulling the hair; in order to make the curve of the curls larger and smoother, the hair needs to be pulled to the opposite direction of the curls.

STEP 4

同样做出 S 形发卷，摆放在顶区与后区的衔接处。

Create S-shaped curls likewise to be placed on the area where the crown and the back sections connect.

STEP 5

将侧面的发丝提拉倒梳后做卷，与顶区发卷衔接。

After pulling and backcombing the hair of the sides, make curls to link with the curls on the crown.

STEP 6

将刘海的发根部分向上梳理，用定型夹固定。发根部分要蓬松，发丝表面要光滑。

Comb the hair roots of the bangs upward and secure the hair roots with a style-setting clip. The hair at the roots should be fluffy and the surface of the hair should be smooth.

STEP 7

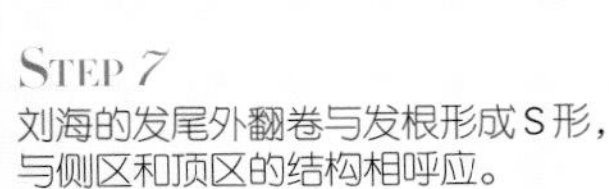

刘海的发尾外翻卷与发根形成 S 形，与侧区和顶区的结构相呼应。

The outward curls of the bangs and the hair at the roots should form an S-shape to match with the crown and side structure.

STEP 8

选择果绿色的可爱花冠，摆放在发卷结构之间。

Place a lovely, fruit green flower crown at the connection area of structure.

整体造型利用卷发的随意多变性处理，蓬松自然，随意收拢，堆发的造型设计更显新娘的活泼个性。

This style takes advantage of the casualness and changefulness of curls. The structure is fluffy and natural. The design of piled-up hair highlights the bride's lively personality.

造型步骤图 STYLING STEP GRAPH

STEP 1
将头发烫卷，将刘海梳出自然随意的形状后固定。
Perm the hair to curls, comb the bangs to natural and casual shape and set the bangs.

STEP 2
将左侧区的头发向顶区收拢，注意将发梢抓出层次。
Gather the hair of the left side to the crown; pay special attention to hand-creating layers in the hair tips.

STEP 3
以同样的手法处理另外一侧。
Treat the hair of the other side in the same way.

STEP 4
用梳子尾端将外翻的发卷向头皮处按压、固定。
Press the outward curls with the end of a comb, press and secure them to the scalp.

STEP 5
将固定好的头发的发梢层次处理蓬松。
Treat the secured layers of the hair tips to fluffiness.

STEP 6
将刘海发丝表面处理顺滑自然，选择花材，佩戴于头发衔接处。
Treat the surface of the bangs hair to make the bangs look smooth and natural; place hairdress flower at the connection area.

整体造型以卷发的发梢设计为主造型结构简洁随意，花冠与绢花的配合更显新娘的婀娜多姿。

This style focuses on the hair tips design of curly hair. The structure is simple and casual. The use of the flower crown and the silk flower highlights the bride's graceful figure.

STYLING 造型步骤图

STEP GRAPH

STEP 1

将所有头发梳顺后烫卷，分出前、后区，在前区基础上分出刘海区，横向取发片倒梳发根。

Comb all the hair smooth and perm the hair into curls, divide the hair into front and back sections; in the front section，separate the bangs and take hair crosswise strand by strand to backcomb at the roots.

STEP 2

将右侧区的头发向后收拢，将发梢留出，与刘海区头发合并。

Fold the right-side hair backward; leave out the hair tips to cover the bangs.

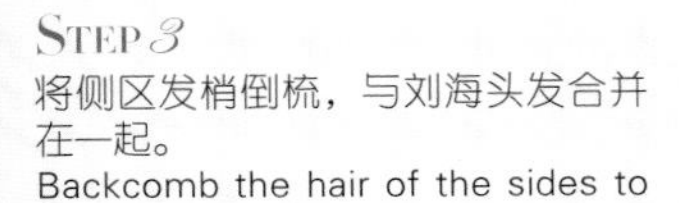

STEP 3

将侧区发梢倒梳，与刘海头发合并在一起。

Backcomb the hair of the sides to cover the bangs.

STEP 4

将后区头发向上收拢，发梢与刘海区合并。

Fold the back hair upward and make the hair tips cover the bangs area.

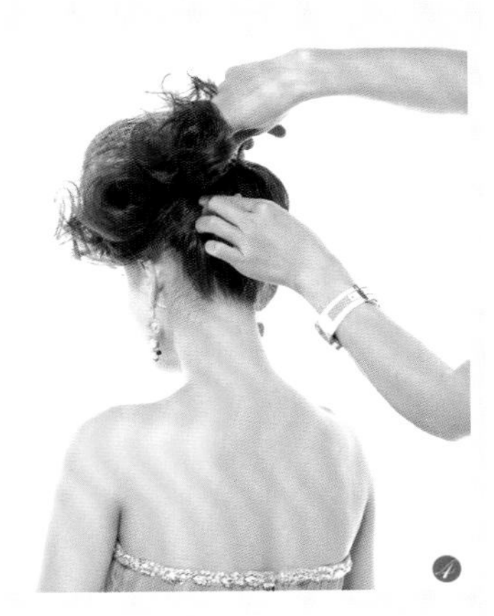

STEP 5

调整好所有头发的层次。

Adjust all the hair layers.

STEP 6

调整好刘海区外侧头发的层次，尽可能避免有空隙。

Adjust the side hair of the bangs area and try to avoid trace of connection.

STEP 7

选择嫩绿色花冠，佩戴于侧区位置。

Place a pale green flower crown on the side.

STEP 8

选择同色系绢花，衔接花冠与发型结构。

Place a silk flower of the same color at where the crown and hair structure meet.

CANDY COLORS

浪漫缤纷的糖果色
Romantic
rainbow colors of
candy

整体造型以高耸的刘海为主，后区头发自然蓬松，高贵而典雅。
This style highlights elegant bangs and the back hair is naturally fluffy, presenting an aura of nobleness and elegance.

造型步骤图 STYLING STEP GRAPH

STEP 1

将所有头发用电卷棒烫卷，并分成刘海区和后区，将左、右侧区的头发分区固定于顶区位置。

Use a curling wand to perm the hair; divide the hair into two sections-the bangs and the back section, and fix the left-side and right-side hair separately on the crown.

STEP 2

为保证左、右侧区的层次，可以将侧区头发分层固定于顶区。

To ensure clear layers of the hair on the right and left, the hair on the sides can be secured into layers on the crown.

STEP 3

将固定在顶区的头发的发梢层次处理好。

Properly treat the layers of the hair tips secured on the crown.

STEP 4

将刘海区的头发向前提拉，倒梳发根，使其形成向前凸起的刘海。

Pull the bangs forward and backcomb the hair at the roots to form flip-up bangs.

STEP 5

将刘海发丝与后区头发衔接好，整理出后区头发的层次。

Connect the bangs with the hair of the back section and sort out the layers of the back section.

STEP 6

以单侧的佩戴方式点缀花材。

Adorn one side of the hairdo with flowers.

PASSION-EVOKING SAPPHIRE BLUE

宝蓝色的激情碰撞

整体造型利用刘海与后区发髻的不对称式设计，凸显新娘的独特之美，鬓边的红玫瑰与黑网纱配合，更增添了造型的优雅与神秘。

This style adopts an unsymmetrical display of the bangs and the back bun to highlight the unique beauty of the structure. The match of red rose and black mesh at the side hair line adds to the elegance and mystery.

STYLING 造型步骤图 STEP GRAPH

STEP 1

将头发分为前、后区，将后区头发以皮筋固定，将刘海倒梳后固定，使其形成大结构的发包。

Divide the hair into front and back sections; secure the back section with a rubber band; backcomb and secure the bangs to form a big bun.

STEP 2

从刘海后面横向分发片，使发梢在额头前面形成弧形层次。

Divide the hair of the back of bangs crosswise into strands; create layered curves at the forehead with hair tips.

STEP 3

选择黑色纱网及花材在侧区佩戴，注意衔接处的弧度。

Select black mesh and flowers to wear on the side, pay attention to make a curving at the connection area.

STEP 4

将后区头发横向分发片倒梳，向上以空心卷挽起并固定，形成发髻。

Divide the hair of the back section crosswise into strands and backcomb, roll up the hair as a hollow roll to be secured, so as to create a bun.

STEP 5

将后区发髻的发梢调整出所需层次并固定。

Adjust the hair tips of the back bun into required layers and secure the hair.

RED AND BLACK
红与黑

整体造型以简洁的盘发设计为主，造型重点在后区发髻位置，层次丰富，更加彰显新娘独特而简约的魅力。

This structure features a simple updo design. The focus of the design is at the back bun. This style has rich layers, fully presenting the uniqueness and modesty of the bride.

STYLING 造型步骤图

STEP GRAPH

STEP 1

将所有头发分片，以大号电卷棒烫卷，分出刘海区、侧区及后区的头发，分发片梳顺，用皮筋在发中位置固定。

Perm the hair with a large-size curling wand; separate the hair into the bangs, the front and the back sections, comb the hair smooth; secure the hair with rubber bands at the middle of the hair.

STEP 2

将侧区中间固定的两片头发合拢，再次用皮筋固定，向内反向翻转，将发梢向下挑出，固定于后区，形成饱满的层次。

Combine the two strands of hair, which has been secured in the middle, and secure the hair with rubber bands one more time; turn the hair reversely and inward, flip out the hair tips to be secured on the back，so as to form rich layers.

STEP 3

将后区用皮筋固定的发片以内扣的形式与顶后区重合，发梢留出备用。以同样的方式处理左、右侧区的头发，衔接固定于后区头发空隙处。

Combine the secured hair strands in the back with the hair of the back crown, flipping the hair inward and leave out the hair tips for further treatment; treat the hair of the left and right sides in the same manner to be connected and secured at the back with no hair.

STEP 4

将顶区发片以同样的方式梳顺，中间以皮筋固定，向后区连接。

Comb the crown hair strands smooth in the same manner, secure the middle with rubber bands, and connect the hair with the back section.

STEP 5

将所有发梢部分的头发分别以空心卷的形式摆放于后区头发空缺处并固定。

Place all the hair tips at the blank area of the back in the form of hollow rolls; secure the hair tip.

STYLING 造型步骤图

STEP GRAPH

STEP 1

将所有头发梳顺，选择大号电卷棒烫卷。

Comb all the hair smooth, and perm the hair with a large-size curling wand.

STEP 2

从刘海区横向分发片倒梳，将刘海发根处理蓬松。

Divide the hair of the bangs section crosswise into strands; backcomb the bangs at the roots to make the hair fluffy.

STEP 3

从右侧开始按外翻卷的弧度自然向后翻卷，将头发固定，使侧区边缘头发自然垂下。

Start from the right section, naturally curl the hair backward according to the curve of outward rolls; secure the hair and let the hair at the edge of the section fall naturally.

STEP 4

将后区多余的头发横向分发片倒梳。

Divide the leftover hair at the back section crosswise into strands; backcomb the hair strand by strand.

STEP 5

将后区倒梳后的头发向上翻卷并固定，形成蓬松的发髻结构。

Flip up and secure the backcombed hair to form a fluffy bun structure.

STEP 6

将左侧头发同样收拢，倒梳发根。

Gather the hair on the left side in the same manner, and backcomb the hair at the roots.

STEP 7

以外翻形式固定于左侧区。

Flip up and secure the hair on the left side.

STEP 8

选择发卡，在两侧固定，点缀发型。

Place hairpins on both sides for fastening and decoration.

整体造型以蓬松的卷发处理出整体效果，造型结构简洁随意，蓬松的卷发更显出新娘的慵懒之美。

This style features fluffy curls. The structure is simple and casual. The fluffy curls add sluggishness to the bride's beauty.

整体造型以蓬松的卷发为设计重点，简洁而随意，更显新娘的简约气质。

This style features fluffy curly hair. It is simple and casual, adding an aura of simplicity to the bride.

STYLING 造型步骤图
STEP GRAPH

STEP 1
将所有头发烫卷，将刘海横向分发片倒梳。
Perm all the hair, divide the hair at the bangs crosswise into strands and backcomb the hair.

STEP 2
用宽鬃毛梳将头发发根向一侧梳理蓬松，将其处理出自然干净的发丝纹路。
Comb the hair at the roots with a broad bristle comb to make the hair fluffy and the hairline natural and clean.

STEP 3
从左侧开始将头发向右侧收拢，固定于后区位置。
Starting from the left side, gather the hair to the right side to be secured on the back section.

STEP 4
选择湖蓝色发卡固定在一侧。
Place a sapphire color hairpin on one side.

STEP 5
用宽鬃毛梳将垂发梳理得蓬松自然，使其垂于一侧。
Comb the hanging hair fluffy and natural with a broad bristle comb, and let the hair fall at one side.

STYLING 造型步骤图 STEP GRAPH

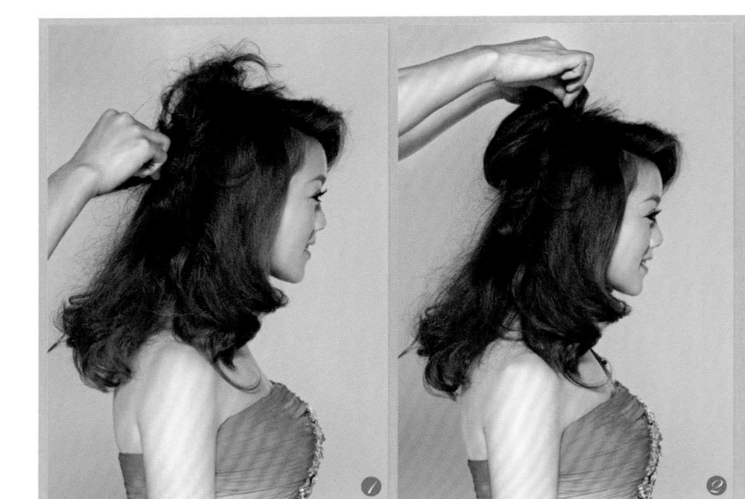

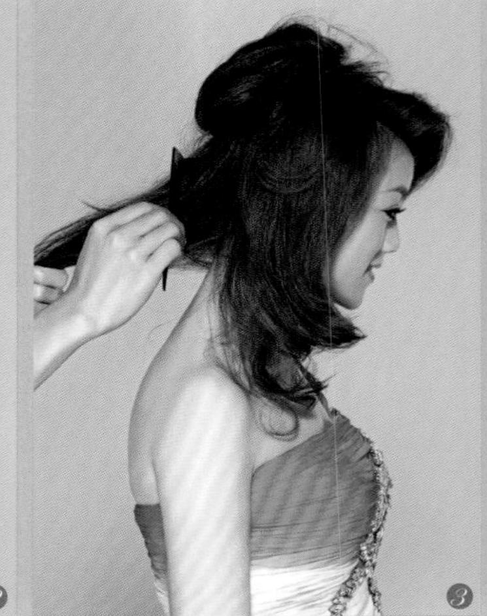

STEP 1

将所有头发梳顺后烫卷，从后区开始横向分发片，向顶区提拉发梢倒梳。

Comb the hair smooth and perm all the hair; start from the back section, divide the hair into strands crosswise; backcomb the hair while pulling the hair tips toward the crown.

STEP 2

将倒梳后的头发向顶区固定，使其形成一个发包。

Secure the backcombed hair on the crown to form a bun.

STEP 3

将后区最后一片头发向下提拉倒梳。

Backcomb the last strand of hair of the back section while pulling the strand downward.

STEP 4

将头发表面梳理光滑，向顶区发包固定，使其形成一个大的结构。

Comb the surface of hair sleek and secure the hair toward the bun on the crown so as to form a big structure.

STEP 5

取侧区头发，横向分发片倒梳，向顶区固定，与后区头发衔接。

Divide the hair of one side crosswise into strands and backcomb the hair strand by strand to be secured on the crown so as to link with the back hair.

STEP 6

从刘海区横向分发片倒梳，向上将表面梳理光滑，固定于顶区。

Divide the hair of the bangs crosswise into strands and backcomb; comb the hair upward to make the surface sleek; secure the hair on the crown.

STEP 7

将另外一侧的头发横向倒梳，向顶区固定，注意刘海区及侧区的衔接一定要自然。

Backcomb the hair of the other side crosswise; secure the hair on the crown; pay attention to make natural connection between the bangs section and the side section.

STEP 8

将右侧区发梢留出，利用头发自然的卷曲度在刘海区与侧区衔接处做一个蓬松的弧旋卷。

Flip out the hair tips of the right side, and use the natural curve of the hair to make a fluffy spiral at the connection area between the bangs and the side.

整体造型以简洁的盘发为主，大结构的盘发更显新娘独特的古典韵味。
This style features a simple updo structure. The big structure of the updo adds the classical aura to the bride.

特色服装造型设计解析

SPECIAL OUTFITS

DESIGNS AND ANALYSIS

整体造型以凤冠设计为重点，红色盖头造型更加突显出清代新娘的喜庆气氛。
This style highlights the phoenix crown. When covered with a red veil, the bride has a festival aura of a Qing Dynasty wedding.

STYLING 造型步骤图

STEP GRAPH

STEP 1

将所有头发梳顺，向后区固定，选择发片，以对称的形式固定于左、右鬓边两侧。

Comb all the hair smooth, secure the hair to the back section, and fasten wig pieces at the hairline of both sides symmetrically.

STEP 2

选择直发发片固定于顶后区，使其分别自然垂于两侧。选择发片修饰额头。

Fasten a straight hair wig on the back of the crown section and let them fall naturally on both sides; select wig pieces to adorn the forehead.

STEP 3

在后区固定大型蝴蝶结发包，注意使蝴蝶结发包一侧尽可能向上，形成发型高度。

Fasten a large-size bowknot bun on the back section, and make sure to place one side of the bowknot bun as upward as possible to form the height of the hairdo.

STEP 4

选择流苏凤钗点缀在造型两侧，在额头位置点缀珠花。

Select tassels and a phoenix hairpin to adorn both sides of the hairdo, then adorn the forehead with beaded flowers.

STEP 5

选择大型凤冠佩戴于顶区位置。

Place a large phoenix crown on the crown section.

整体造型利用不对称的设计手法，珠花与凤钗的点缀将整个外围轮廓修饰饱满，凸显古典造型之美。
This style adopts an asymmetrical design. The beaded flower and phoenix hairpin add to the fullness of the peripheral lines to highlight classical beauty.

STYLING 造型步骤图

STEP GRAPH

STEP 1

将所有真发梳顺，固定于后区，选择 U 形假发片修饰额头。

Comb all the hair smooth and secure on the back section, adorn the forehead with a U-shape wig.

STEP 2

选择小型蝴蝶结发片，在右侧摆放固定，形成不对称式设计，用细发辫修饰连接蝴蝶结发片。

Fasten a small-size bowknot wig piece on the right side so as to form an unsymmetrical design; use thin braids to modify and connect the bowknot wig piece.

STEP 3

选择"燕翅"假发，斜向固定于顶区，并用粗发辫将整个外围轮廓连接固定，后区用蝴蝶结发包填充发型的不足之处。

Fasten a dovetail wig obliquely on the crown, and use a thick braid to connect and secure the entire peripherals; use a bowknot bun to fill up the deficiency of the back section.

STEP 4

选择珠花与凤钗在头发衔接处佩戴。

Select beaded flower and a phoenix hairpin to adorn the connection areas of the hairdo structure.

STEP 5

在刘海区一侧固定彩色珠花，衔接刘海与侧区的位置。

Fasten a colorful beaded flower on one side of the bangs section to connect the bangs and the wig of the side section.

整体造型结构简洁饱满，珠花与牡丹的点缀更能突显出唐装造型的雍容华贵。

This style has a simple and full structure. The use of beaded flower and peony highlights the elegance and luxury of Tang Dynasty outfits.

STYLING 造型步骤图
STEP GRAPH

STEP 1

将所有头发梳顺，固定于后区，选择粗发辫，将头部外围轮廓修饰圆润。选取牛角包在一侧固定。

Comb all the hair smooth to be secured on the back section, use a thick braid to modify the whole peripherals to fullness, and fasten a croissant bun on one side.

STEP 2

在牛角包前选择中型蝴蝶结发包遮挡，用发片缠绕于牛角包表面，增加发量。

Place a medium bowknot bun in front of the croissant bun to cover the latter, roll hair pieces around the surface of the croissant bun to increase hair volume.

STEP 3

将发片另外一侧梳顺，在中间部分用皮筋固定，使其形成饱满的结构，固定于右侧区。将剩余头发以发髻的形式固定于后区。

Comb the other side of the hair pieces, secure the middle part with a rubber band to form a full structure to be secured on the right side, and then secure the remaining hair at the back section in a bun.

STEP 4

选择粉色牡丹花佩戴于顶区结构处，结构衔接处以凤钗点缀。

Place a pink peony on the crown, and decorate the connection area of the hair structure with a phoenix hairpin.

STEP 5

注意侧区的发型饱满度及弧度。

Pay attention to the fullness and curve at the side section.

整体造型以对称饱满的结构设计为重点，凤钗与红色牡丹的运用更加彰显出唐装造型的尊贵气质。

This style features a symmetrical and full structure. The use of phoenix pins and red peony further presents the nobleness of Tang Dynasty outfits.

STYLING 造型步骤图 STEP GRAPH

STEP 1

将所有头发梳顺，固定于后区，刘海处选择U形发片固定，修饰额头位置。选择中间固定的发片固定于顶区，从一侧开始处理成对称式发髻。

Comb all the hair smooth and secure the hair on the back section; select a U-shape wig to be fastened on the bangs; modify the forehead area; select a wig with fixed middle section to be fixed on the crown; starting from one side, create a symmetrical bun.

STEP 2

将蝴蝶结发包竖向固定于顶区，在蝴蝶结发包上用牛角包修饰固定，后区以蝴蝶结发包作为整个顶区造型的支撑。

Fasten a bowknot bun on the crown lengthwise, and fasten a croissant bun on the bowknot bun; for the back section, use the bowknot bun as the support of the whole crown structure.

STEP 3

选择红色牡丹花材佩戴于顶区。

Place red peony flower on the crown.

STEP 4

左右空缺处选择金色凤钗，以对称的方式佩戴，顶区选择朱红色发钗点缀，后区用凤冠修饰。

Place gold phoenix hairpins symmetrically at the blank areas of the left and right sides, adorn the crown with a scarlet hairpin; adorn the back section with a phoenix crown.

STEP 5

选择流苏修饰鬓边两侧。

Select tassels to adorn the hairlines of the sides.

单侧的发髻摆放一改古装对称的发型设计，在古典中融入时尚感，珠花饰品和大朵牡丹增加了人物的高贵气质。
The lopsided placement of the bun goes against the symmetric design of ancient outfits so as to add fashionable elements to the classical style. The use of beaded flowers and big peony add noble aura to the hairdo.

STEP 1

将头发全部收拢至头顶扎马尾，将发尾缠绕固定，在右侧耳后上方固定牛角包。

Gather all the hair to the crown to be tied into a ponytail, coil and secure the hair tips, fasten a croissant bun on the upside of the back at the right ear.

STEP 3

后区用发包做低髻，尽量使其饱满圆润。

Wrap the back hair into a low bun, which should be as full and round as possible.

STYLING STEP GRAPH 造型步骤图

STEP 2

用发包将牛角包和发型之间的空隙添满。

Use a hair wrap to fill up the blank between the croissant bun and the other parts of the structure.

STEP 4

选择大朵牡丹在顶区位置固定，以修饰整体发型的饱满度；珠花以不对称的方式点缀，增加造型的层次。

Select a big peony to be fastened on the crown so as to add to the fullness of the overall structure. Adorn the hairdo with beaded flowers in an asymmetry manner, so as to add to the layers of the structure.

此款发型简单易操作，避免了过多假发带来的厚重感，而是用配饰来体现古韵，发钗和流苏耳环的巧妙搭配让华丽感更加突出。

This style is simple and easy to make, avoiding the heaviness due to excessive use of wigs. It uses accessories to create an aura of ancient outfits. The ingenious combination of hairpins with tasseled earrings highlights the aura of luxurious beauty.

STEP 1

将头发全部扎马尾，在顶区固定，选择细发辫，以对称的形式固定于顶区，将发辫以S形组合成对称的发片修饰额头，发辫发尾自然垂下。在额头中间摆放假刘海。

Tie all the hair into a ponytail to be secured on the crown, and select thin braids to be fastened on the crown symmetrically, use braids to make S-shaped symmetrical hair pieces to adorn the forehead, and let the end of the braids fall naturally; place a false bang in the middle of the forehead.

STEP 2

在顶区向左、右耳后位置各固定一个牛角包，让发型外轮廓圆润。

Fix a croissant bun on the crown behind both the right and the left ears to make the peripheral lines of structure mellow and full.

STEP 3

后区选择发髻固定，将发钗以对称的形式固定于后区发髻上。

Select a bun to be secured at the back, and select hairpins to be pinned at the back bun symmetrically.

STEP 4

选择整顶花冠固定于顶区，使流苏自然向两侧垂下，顶后区选择珠花以对称的形式点缀。

Secure a full corolla on the crown, let the tassels fall naturally on both sides, and select beaded flowers to adorn the back of the crown symmetrically.

用简洁的手法打造饱满的发型轮廓，华丽的凤冠配饰将唐装雍容华贵的气质体现得淋漓尽致。
This style uses simple techniques to create a hairdo of full structure. The use of luxuriant phoenix crown adds to the grace and luxury of Tang Dynasty outfits.

Step 1
将头发全部收拢至头顶固定。取假发片在顶区固定，以对称的形式设计两侧的发髻，注意发丝表面的光滑度。
Gather all the hair to the crown and secure the hair. Take strands of the wig to be fastened on the crown, create symmetrical buns on two sides; pay attention to the smoothness of the surface of the hairdo.

Step 3
在后上方固定发髻，衔接蝴蝶结假发。
Fasten a bun on the back of the crown to link with the butterfly tie wig.

Step 2
在顶区固定圆形大发包，在发包的后方固定小蝴蝶结假发，做出发型的高度。
Fasten a large round bun on the crown, and fasten a small bowknot wig on the back of the bun so as to create the height of the hairdo.

Step 4
在发型后方结构的空缺处佩戴大朵牡丹花，在结构线处再用小朵花材点缀，在发型前方戴上凤冠。
Place a big peony on the blank area of the back structure，adorn the structural lines with small flowers, and place a phoenix crown at the front of the hairdo.

整体造型以对称式设计为主。珠钗与步摇点缀其中，突出清代女子的秀美与内敛。

This style features a symmetrical design. The tassels and tasseled pins adorn the hairdo, presenting an aura of a graceful and modest female in Qing Dynasty.

STYLING 造型步骤图 STEP GRAPH

STEP 1

将头发梳顺，分出前、后区，将后区头发固定，选择直发发片固定于后区，并用发辫以8字形固定于发中位置。

Comb the hair smooth, divide the hair into front and back sections, secure the hair at the back section, select a straight hair pin to be fastened to the back section, fasten a braid in 8-shape in the middle.

STEP 2

将所有前区头发梳顺后固定于顶区，选择牛角包在顶区固定。

Comb all the front hair smooth to be secured on the crown, and fasten a croissant bun to the crown.

STEP 3

选择蝴蝶结发片，遮挡牛角包与后区发片的衔接位置。

Use bowknot wig piece to cover the connection area between the croissant bun and the back hair.

STEP 4

选择小号蝴蝶结发片，竖向固定于牛角包中央的位置，形成顶区结构的高度。

Select small-size bowknot wig piece to be fastened on the middle of the croissant bun lengthwise, so as to form the height of the crown structure.

STEP 5

选择流苏步摇佩戴在一侧，用珠花点缀整个造型。

Place tassels and a tasseled pin on one side, and use beaded flowers to adorn the overall structure.

整体造型简洁大方，对称式的设计一直是古装造型的经典形式，流苏珠钗更赋予造型生命力，是古典造型的首选。

This style is based on a simple structure. The symmetrical design has always been a favorite in the ancient costume styling. The tassels and phoenix pins will add vitality to the hairdo and are perfect for classical styling.

STYLING 造型步骤图 STEP GRAPH

STEP 1

将所有头发梳顺，分出前、后区，将后区头发中分，选择"燕翅"固定于中间，在后区两侧倒梳头发，以交叉包发固定好"燕翅"发髻。

Comb all the hair smooth, divide the hair into front and back sections, part the back section in the middle, fasten a dovetail bun in the middle, backcomb the side hair at the back section, and secure the dovetail bun in a crossing wrap.

STEP 2

将前区中分，以对称的手法固定头发，选择蝴蝶结发包在顶区位置固定，以此增加发型高度，衔接前、后区结构。

Part the front section in the middle, secure the hair symmetrically, fasten a bowknot bun on the crown to add height and connect the front structure with back structure.

STEP 3

将所有头发处理干净，将发丝固定整齐，使真假发衔接自然。

Trim all the hair neat, secure the hair strands tidy, and link the real and false hair strands naturally.

STEP 4

选择发片，在前区位置固定，将发片两侧以空心卷设计，选择流苏凤钗，以对称的方式佩戴于蝴蝶结发包上。

Fasten a hair wig on the front section, create hollow rolls on both sides of the hair strands, and select tassels and a phoenix hairpin to adorn the bowknot bun in a pile-up form.

STEP 5

在前区正中位置点缀珠花，在左右两侧也点缀珠花。

Adorn the center of the front section and both sides with beaded flowers.

整体造型以对称设计手法为主，珠花与蝴蝶的点缀，更加凸显出晚清时期小家碧玉的形象。

This style features a symmetrical design. The use of beaded flowers and butterfly adds an aura of a modest girl in late Qing Dynasty.

STYLING 造型步骤图 STEP GRAPH

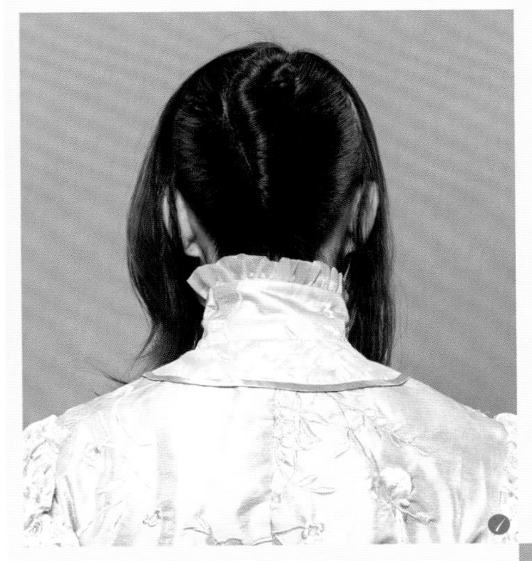

STEP 1
将头发梳顺，分出前、后区，将后区包发固定，将前区中分。
Comb the hair smooth, divide the hair into front and back sections, secure the back section with a wrap, and part the front section in the middle.

STEP 2
将前区头发中分，向两侧以包发设计。
After parting the front hair in the middle, wrap the hair towards both sides.

STEP 3
选择假发片固定于顶区位置，将发片均分，编成松散干净的三股辫备用。
Select a wig piece to be fastened on the crown, divide the wig piece evenly, and create loose and neat three-strand braid for further use.

STEP 4
将两条三股辫固定于后区位置，并用细发辫修饰后区两侧，前边选用长条发包对称固定。
Secure the three-strand braids at the back section, select thin braids to modify both sides of the back section, select long buns to be fastened symmetrically at the front.

STEP 5
选择珠花与流苏点缀于对称的长条发包尾端，选择蝴蝶及珠花点缀在造型中间。
Select beaded flowers and tassels to adorn the ends of the long buns; select butterfly and beaded flowers to decorate the structure.

整体造型以对称式的手法进行处理，发型结构简洁干净，表现出女性端庄秀美的气息。
This hairstyle is symmetrical. The structure is simple and neat. The hairdo shows dignified feminine beauty.

STYLING 造型步骤图 STEP GRAPH

STEP 1

将头发梳顺，分出前区、后区、顶区和侧区。将刘海区和后区均以 5: 5 分区，将顶区头发三股编辫后盘绕固定。

Comb the hair smooth, separate the hair into front, back, crown and side sections. Both the bangs and the back section are parted into 5:5, braid the crown hair into a three-strand braid and coil and secure the braid.

STEP 2

把燕尾式假发固定在后枕骨处，将后区头发左右交叉式包住燕尾假发。

Fasten a dovetail wig on the lower back, and cross the hair of back section to wrap the dovetail wig.

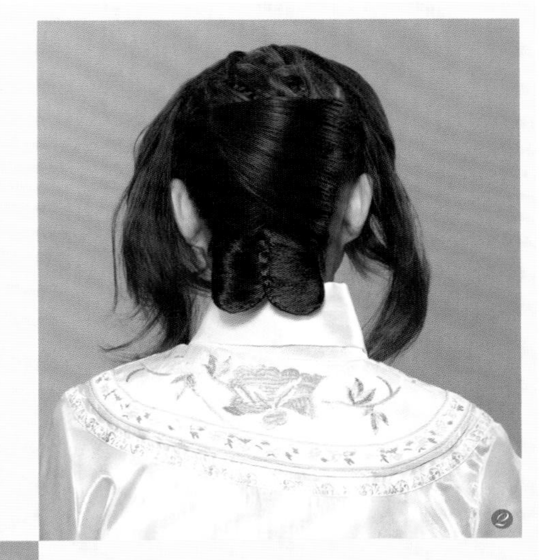

STEP 3

将侧区发片内侧倒梳，向后区包发衔接，将刘海区的头发以手推波纹的方式处理，固定在侧区发片上。

After trimming the inside of the hair strands of the side sections, wrap the hair backward to connect the back section, and hand-push the bangs into waves to be secured over the hair of the side section.

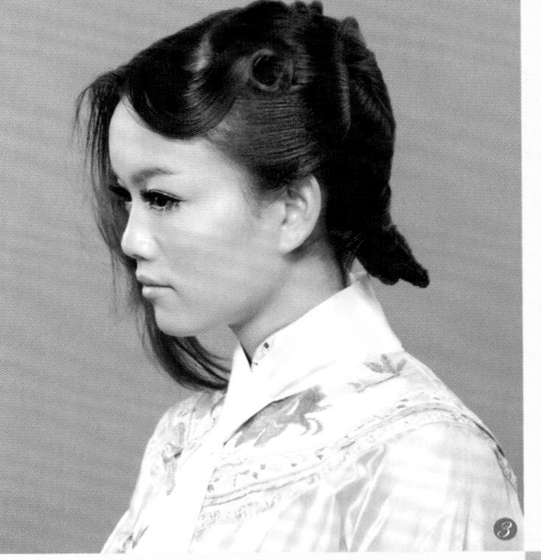

STEP 4

用发排在顶区进行垫发处理，增加发型的高度，两侧位置用假发卷塑造造型结构，要求左右对称。

Pad the crown with a hair row to increase the height of the structure, and select wig rolls to shape the structure of both sides, which shall be symmetrical.

STEP 5

选择发钗和点翠款饰品，在发型的两侧和后区进行点缀。

Select hairpins and jade ornaments to adorn both sides and back section.

不对称的发型设计与直发假发片结合运用，使人物显得清秀脱俗，避免了盘发使人年龄感偏大的问题，蝴蝶饰品让发型更有少女气息。

This style combines asymmetrical design and the use of straight hair. The hairdo presents a graceful and refined aura and avoids the updo's aura of agedness. The butterfly ornaments make the hairdo look more youthful.

STYLING 造型步骤图 STEP GRAPH

STEP 1

将头发分出刘海区，用鸭嘴夹固定，将其余头发全部扎马尾至顶区固定，选用假发片在顶区右侧固定。

Separate out and secure the bangs section with a duck-bill clip, tie up all the remaining hair into a ponytail to be fastened on the crown; fasten a wig piece on the right side of the crown.

STEP 2

另一侧以同样的方式固定假发片，将刘海区头发表面梳理干净，向顶区固定，做出有高度的刘海。

Fasten the wig piece on the other side in the same manner, and then comb the surface of the bangs to be secured to the crown, so as to create bangs with height.

STEP 3

在后区选用直发片固定，将假发片梳顺后用发带扎紧，系蝴蝶结装饰。

Select a straight hair wig piece to be fastened on the back section; after combing the wig smooth, tie the wig hair with a bowknot for decoration.

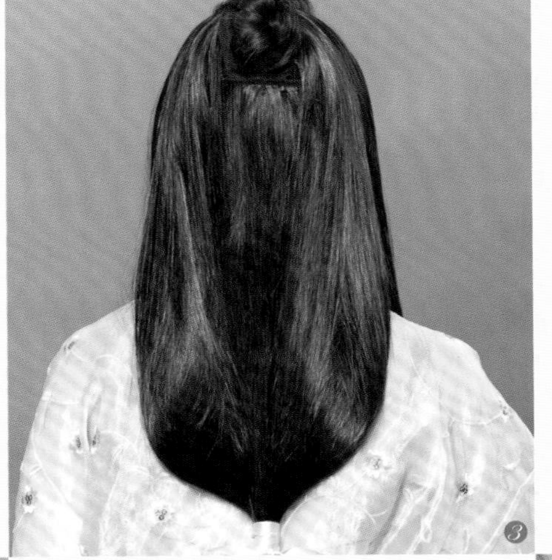

STEP 4

选择古装用假发髻，在顶区固定，打造发型的高度。

Select a wig bun for ancient costume to be fastened on the crown so as to create the height of the hairdo.

STEP 5

选用蝴蝶结假发片，斜向摆放于顶侧区位置并固定。用发片修饰顶区与侧区，将蝴蝶饰品及珠花佩戴于侧区位置，以发钗步摇点缀。

Select a bowknot wig piece to be fastened obliquely on the side of the crown, take strands of hair to modify the crown and side sections; select butterfly ornaments and beaded flowers to adorn the side sections, and use hairpins and tassels for decoration.

回眸一笑

百媚生。

STYLING STEP GRAPH

造型步骤图

STEP 1

将头发分成前、后区，将后区头发扎马尾，固定在正后方，将前区中分。

Divide the hair into front and back sections, secure the hair into a ponytail on the back, and part the front section in the middle.

STEP 2

将后区发尾固定，选择发髻固定在枕骨处，形成向下的饱满弧度，将刘海梳理干净，做包发处理。

Secure the hair tips of the back section, and select a bun to be fastened on the lower back, so as to form a full downward back curve. Comb the bangs and create a wrap.

STEP 3

将刘海做中分包发处理， 盖住部分耳朵，来体现温柔的感觉。

Part the bangs at the middle and make wraps; cover part of the hair to give a feeling of gentleness.

STEP 4

选择发棒在顶区做发髻设计，选择发包做顶区弧度，用精美的珠花点缀，衔接发型结构。

Select a hair stick to make a bun on the crown, and make a wrap on the crown to create a proper curve; adorn the structure with exquisite beaded flowers to match the hairstyle.

STEP 5

选择大朵珠花，佩戴在后区发髻结构的衔接处。

Select a large beaded flower to be placed at the connection area of the back bun structure.

此款造型主要突出新娘贤德的古典的气质。发型结构简单，恰到好处地突出了那个年代的造型特点。

This style mainly highlights the virtuous and classical disposition of the bride. The simple hairstyle appropriately reflects the characteristics of the hairstyle of old days.

STEP 1

将头发梳顺，分出前、后区，将后区头发在枕骨下方梳成一个发髻结构，前区头发以5:5分区，在耳后发髻处进行固定。

Comb the hair smooth, divide the hair into front and back sections; comb the hair of the lower back into a bun structure; part the front hair to 5:5 and secure the hair of the bun behind the ears.

STEP 3

把单头假发排均分为两份，分别编成发辫固定在顶区的位置。用青花瓷布艺固定在上方。

Equally divide a single-end wig into two parts to make a braid for each part; secure the braids on the crown. Fasten a fabric art piece of blue-and-white porcelain pattern on the crown.

STEP 2

先在后枕骨处固定一个假发排，进行垫发处理，再在假发排上端固定一个蝴蝶结发包，最后在前区分区线处摆放蝴蝶结假发。

Fasten a wig row behind the lower back to pad the hair, fasten a bowknot wig on the upper section of the wig row, and place a wig bowknot at the center of the front parting line.

STEP 4

收起发辫，在整个造型的外轮廓上进行固定。选择具有中国特色的镶嵌式银饰品和发钗，在后区、前区和两侧的位置佩戴。

Secure the braids on the peripherals of the structure; select mosaic silver ornaments and hair pins with Chinese traditional design to adorn the back, front and both sides of the hairdo.

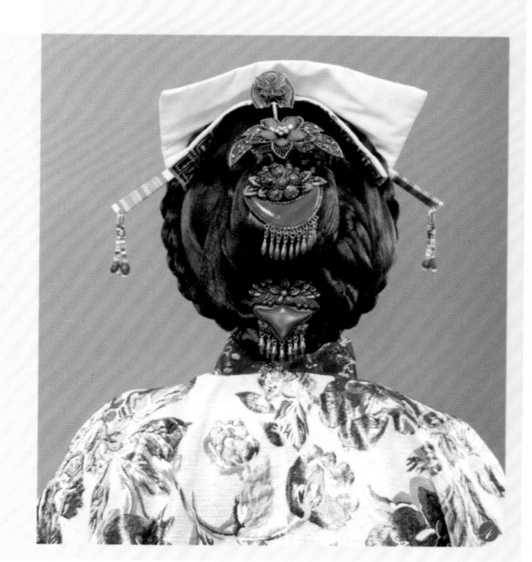

整体造型技法简洁，轮廓圆润饱满，饰品运用主次分明，表现出女性成熟大方的一面。

The technique used for creating the structure is simple and neat. The outline is rich and full, and the arrangement of the ornaments prioritizes their roles. This style presents the mature and generous side of feminine disposition.

整体造型运用了不同的假发和不对称式的结构处理方法，在倾髻的基础上加入了修饰的手法。饰品的运用表现出了喜庆妩媚的女性魅力。

This style features the use of different wigs and an asymmetric structure. Adding to the technique for oblique buns, the hairdo uses decoration skills. The ornaments show festival and enchanting feminine charms.

STYLING STEP GRAPH 造型步骤图

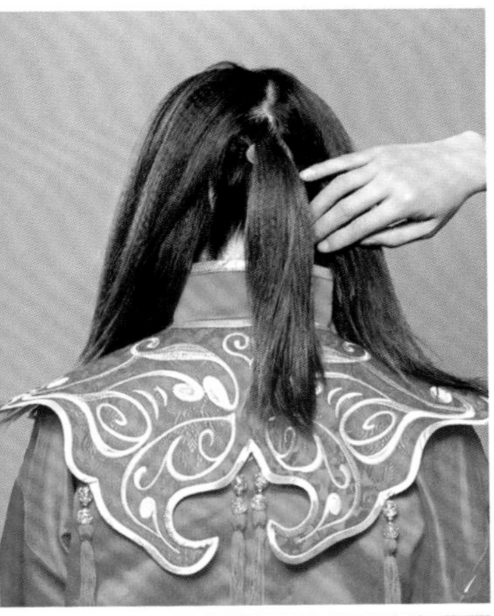

STEP 1

将头发梳顺。分出前、后区，将后区发际线至枕骨上线的头发以三角形分区扎成发髻备用。

Comb the hair smooth. Divide the hair into front and back sections; tie a bun with the hair of the triangular area formed by the back neck line and the middle of the back section.

STEP 2

把燕尾式假发固定在后发髻的位置，将其用后区余发交叉包住。

Fasten a dovetailed wig on the back bun, and wrap the wig by crossing the remainder hair on the back section.

STEP 3

前区以 1:9 分出大刘海区，将侧区发片向后区做包发处理。

Part the front section into a big bangs section at the ratio 1:9; wrap the side hair toward to the back section into a bun.

STEP 4

右侧选择蝴蝶结式假发，左侧选择蝴蝶包式的假发，在顶区分缝线处进行轮廓填充。

Use a bowknot wig for the right side and a bowknot bun wig for the left side, pad up the peripherals at the parting line on the crown.

STEP 5

造型顶部选择假发排和燕尾式假发，进行造型轮廓的提升。选择不同材质的玫红色饰品，以不对称的方式进行点缀。

Select a wig row and a dovetailed wig to shape up the crown; select magenta decorations of different texture to adorn the hairdo in an asymmetric method.

整体造型利用蓬松的倒梳手法完成，刘海区与侧区合为一体，更加突显出旗袍造型的优雅妩媚。

This style features fluffily backcombed hair. The bangs and side sections merge into one, fully matching the elegance and charm of cheongsam outfits.

STYLING 造型步骤图
STEP GRAPH

STEP 1
将所有头发用大号电卷棒烫卷，从顶区横向倒梳发片，以空心卷向上固定，将后区剩余的头发以同样的方式横向分发片倒梳，固定于顶区，形成蓬松的饱满结构。
Perm all the hair with a large-size curling wand; start from the crown, backcomb the hair strand by strand crosswise and create hollow rolls to be secured upwards; divide the remaining hair of the back section crosswise in the same way and backcomb the hair to be secured on the crown，so as to form a fluffy full structure.

STEP 2
从侧区竖向分发片倒梳，向后区包发，连接后区。
Start from the side section，divide the hair lengthwise into strands and backcomb; wrap the hair towards the back section to connect the back section.

STEP 3
另外一侧以同样的方式竖向分发片包发，连接后区，将发梢固定于顶区位置。
On the other side, divide and wrap the hair lengthwise in the same way to connect to the back section; secure the hair tips on the crown.

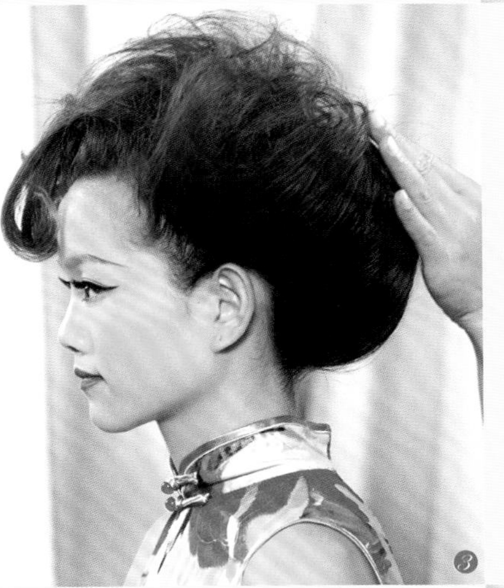

STEP 4
将刘海倒梳蓬松，发梢留出，向一侧覆盖于侧区表面。
Backcomb the bangs fluffy; flip out the hair tips toward one side to cover the surface of the side section.

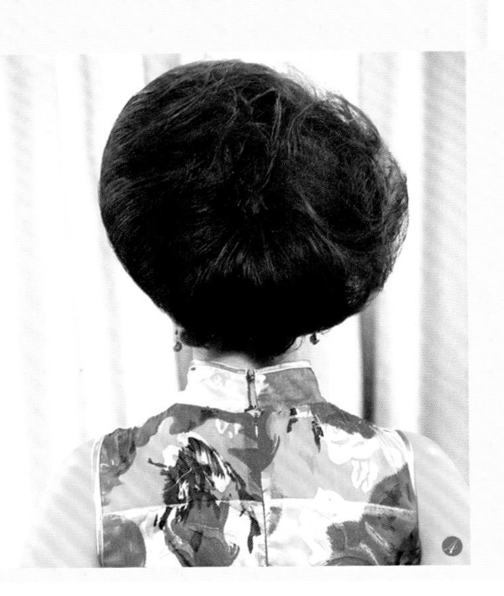

STEP 5
用干胶将饱满的结构固定住。
Set the full structure with hairspray.

SMECTA

U2杯职业摄影师大赛／首席摄影师一年班毕业汇报
杯职业化妆师专修班毕业作品主题发布会
2012黑光·朵莉西雅杯职业化妆师大赛
贺黑光十九华诞
祝母校再创辉煌
颁奖大

京墨光摄影化妆数码培训学校毕业联

黑光教育简介

黑光集团成立于1992年，下设黑光网、黑光人才网、黑光教育、黑光资产管理等多个事业部。

黑光教育，从属于黑光集团，致力于做全球摄影化妆教育第一品牌。

强大的师资阵容、严格的教学质量、良好的口碑与品牌形象建设，使黑光教育受到社会各界的广泛赞誉，成为业界的翘楚。

黑光学子遍布海内外，每年吸引大批业界精英前来进修提升，多年来为业界培养了大批摄影、化妆、数码、管理等专业人才，毕业学子或自主创业，或进入相关企业，多年过去，很多学子现已是行业内的中坚力量。同时，依托黑光人才网和对接的用人企业，黑光集团为学子搭建了一个优秀的交流与工作的平台。

黑光教育每年多次应邀参加国内、国际摄影化妆大型交流活动，并得到了包括中央电视台在内的几十家机构提供的强大的媒体支持。

黑光教育是行业的一个窗口，对业界产生着积极的导向作用，被誉为中国摄影与化妆界的一面旗帜。同时，黑光朝着国际化发展的进程已经启动，黑光已经在美国、日本、韩国、马来西亚、泰国等地进行高端教学合作并取得了巨大成功，为学员开拓了国际化教育的道路。今天的黑光，正以轻快的步伐、行业领跑者的豪气，实现着不断超越的愿景。

黑光教育，愿做最黑的夜中，那最亮的光……

【黑光教育大事记】

1992 年 黑光集团的前身——黑光摄影工作室成立

1998 年 黑光影楼人才培训中心成立

2002 年 编辑出版黑光系列丛书之《影楼策划》《影楼门市》《黑光管理》等

2003 年 黑光被授予"中国美容行业名校"称号

2004 年 黑光被授予"摄影化妆名校"称号
编辑出版管理类系列丛书之《影楼从业人员必读》《影楼经理人必读》

2005 年 黑光孙小平老师荣获"中国人像摄影十杰"称号

2006 年 黑光李泽、黄煌老师均荣获"中国十佳化妆师"称号

2007 年 黑光荣获"2007 美丽中国年度特别成就奖"
黑光荣获"2007 教育杰出贡献奖"
美国 PPA、韩国 PPK 摄影家协会来北京黑光进行学术交流
《黑光影楼数码化妆造型宝典》隆重出版，并荣获 2007 年度美国印制大奖 COM 奖项

2008 年 黑光组织师生为汶川地震灾区捐款，并出资拍下义卖剪纸作品《清明上河图》，奉献企业爱心
黑光参加奥运圣火在天安门广场的传递仪式
黑光被国家文化部相关部门授予"中国美业华表奖——中国美业诚信经营机构"称号

2011 年 黑光李泽老师荣获"资生堂中国化妆大赛全国总冠军"

2012 年 黑光被授予第 8 届中国健康美容信誉联盟荣誉单位
黑光阿喜老师荣获"第 12 届全国影展金奖"及"第 7 届全国人像摄影"提名奖。
黑光集团 20 周年庆典隆重举办
黑光教育国际部首届留学生班成功开办

未来，我们还将继续书写历史……

【联系我们】

企　　业：北京黑光教育咨询有限公司
地　　址：北京市海淀区西四环中路 15 号（邮编 100143）
黑 光 网：www.heiguang.com
黑光人才网：hr.heiguang.com
黑光教育网：school.heiguang.com
企 业 Q Q：8000-19308
招生与图书热线：400-085-8881
求职与就业热线：010-88110848

后记

盛世如约，物换星移。夜已深，独坐灯下，思绪回到 1992 年 3 月 28 日……

那一天，是一个特殊的日子，那一天，是黑光诞生的日子。

今天，黑光迎来了 22 岁华诞。22 年了，其间的坎坷经历、执着奋斗的历程，深刻地铭记在每个黑光人的心间。成长是快乐的，成长也是艰辛的，正如一个咿呀学语的孩子，每天我们对他倾注心血，每天我们也对他饱含希望。我们在殷切期盼中奋斗成长！

黑光是一群年轻人用智慧、心血和汗水浇灌着的一项美丽事业！伴随着黑光的成长，伴随数不清的日日夜夜，他们在艺术事业上取得了一个又一个令人羡慕的成就，他们创造着美、传播着美、收获着美，黑光是他们用勤劳、智慧和汗水浇灌起来的一棵参天大树。

如今，在业界的翘首企盼中，饱含这群年轻人无数心血与智慧的又一结晶《黑光造型 经典造型设计专业教程》与《黑光造型 创意造型设计佳作赏析》面世了！他们又为业界交上了一份令人满意的答卷，同时为黑光献上了一份最独特的生日礼物！

本书的面世，正如同黑光的成长，同样离不开一直以来帮助、关心与支持黑光的各位领导、前辈、好朋友，以及众多媒体、相关单位和个人的支持，在这个特殊的日子里，真诚地感谢你们！

感谢黑光创意团队的主创人员！感谢为此书提出宝贵意见以及参与修改工作的老师们！他们为本书的化妆、摄影、数码设计与修改付出了大量的心血，使得本书能够如期顺利出版，感谢所有战斗在前方与后方的黑光战友们，以及给予他们大力支持的黑光战友的家人们。也感动于多年来我们一起相伴黑光、风雨同行的日子……

今天，黑光担负着授业、解惑的职责，为所有的同行和黑光学子搭起了一个学习与交流的平台。常常会有友人和同学向我诉说成长中的快乐与艰辛。他们曾带着梦想来到北京，来到黑光，现在，他们正慢慢实现着自己的梦想，或在影楼发展得越来越大，或刚刚开了工作室，或当了编辑，办起了刊物，有的已在时尚界小有名气…… 他们遍及世界各地，黑光可谓桃李满天下。每每想到此，心中便倍感欣慰，再辛苦、再累也是值得的！是我们这些可爱的黑光学子，让黑光的美名蜚声海内外，在此我也深深地感谢你们，我可爱的黑光学子们！

面对一双双求知的眼睛，面对来自海内外的莘莘学子，我们便感到肩上更多了一份责任。真诚希望大家都能在奋斗的路上快乐着、成功着。我将永怀一颗感恩之心，为所有的友人和新老同学送上最真诚的问候与祝福！

人生的旅行中，永远期待与大家重逢！多年以后，不管你想不想起，黑光就在那里，不离、不弃……

黑光集团创始人之一